D0684830

ESSENTIAL MATH FOR
PHYSICAL CHEMISTRY

DAVID W BALL

PREFACE

I have been teaching physical chemistry for almost 30 years. I even have a textbook in physical chemistry that's gone into a second edition (which means that the publisher thinks it's not crap). In my humble opinion, the reason why students have so much trouble in physical chemistry, the reason why it's widely considered such a hurdle for chemistry students, is because of the math.

Physical chemistry is based largely on calculus, just like general chemistry is based on algebra. (By the way—general chemistry students have problems with math, too. I have another book that addresses that.) If students aren't that familiar with their calculus, they'll struggle with the chemistry. Again, in my humble opinion, this is confounded by two issues. First, most physical chemistry students have taken calculus, but it was a while ago, sometimes years before they got into their physical chemistry class. It's my guess that the calculus classes they took didn't use chemistry examples when they were presenting the material, so students never got the chance to see how it applies to chemistry (please, math teachers, don't take that as a criticism. I know you have many constituencies to satisfy). Second, in my experience many physical chemistry instructors don't seem to sympathize with the math struggles that the students have. THEY understand the math—"It's easy! I don't understand how students can't get this. And besides, I'm not a math teacher; I expect my students to either know the math they need to know or relearn it on their own." (I actually had a colleague at another institution say something like that to me.) And so the problems continue. At least, the *students'* problems continue…

As with algebra in general chemistry, students going into physical chemistry usually have had calculus but are functionally illiterate in it —they can't apply the math to the physical chemical principles they have to master. So, they do poorly, and end up blaming physical chemistry itself, or their instructor, or their book, and they end up making bumper stickers that bad-mouth the field.

What physical chemistry students need is a review focused on the calculus skills they will need in their physical chemistry class. And that's what I've put together here. It's based on my own experiences in the physical chemistry classroom and the skills that I see my students needing to be successful. It may not cover every calculus- or math-based skill that some textbooks-*cum*-encyclopedias might cover, but it should cover most of the fundamental math skills a physical chemistry student needs. (Besides, why are people using an encyclopedia as a textbook?)

My sincere thanks to my colleague Katelynn Edgehouse for reviewing the manuscript, pointing out errors, and making comments. I appreciate it, Kate!

Any feedback or suggestions are welcome, especially if there's a gaping hole in the coverage. Hopefully the review of math skills presented here is helpful. Good luck!

David W. Ball

December 2018

PART I

DIFFERENTIATION

1

THE SLOPE

A mathematical equation, or a **function**, is an expression that relates two quantities called **variables**. Variables are usually (but not always) represent in italics, like x. Here is an equation that relates two variables x and y:

$$y = 2x + 1$$

This equation means that whatever the numerical value of x is, multiply it by 2 and then add 1 to get the value of y. If x were equal to 3 (written as "$x = 3$"), then we can calculate y as

$$y = 2(3) + 1 = 6 + 1 = 7$$

Different values of x can be used to determine the resulting value of y. We can plot these values on a graph, which typically puts the x value, called the **independent variable**, on the horizontal axis (the "x axis") and the y value, called the **dependent variable**, on the vertical axis (the "y axis"):

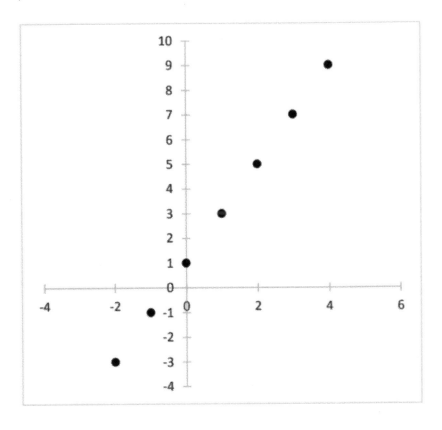

In the graph above, the tick marks on the axes mark the respective values of x and y, and the dots represent the points that satisfy the equation $y = 2x + 1$. To make it easier to see, we can draw a line to connect the dots and then simply remove the dots:

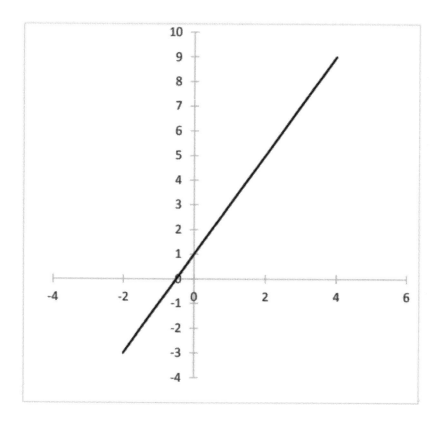

The line is a graph that illustrates the equation $y = 2x + 1$. Other mathematical equations can be graphed similarly.

Graphs like this that show a straight line have two useful properties: a y-intercept and a slope. The **y-intercept** is the value that y has when the plot crosses the y-axis (which occurs when $x = 0$). By looking at the graph, we can see that the plot crosses the y-axis at a value of $y = 1$.

The **slope** is a measure of how fast the function changes its value of y, going from left to right. The fundamental definition of slope is

$$\text{slope} = \frac{\text{change in } y}{\text{change in } x} = \frac{\Delta y}{\Delta x}$$

where in the last expression, the uppercase Greek letter delta, Δ, is commonly used to represent "change." (We will see variations of this later.) Another way to express the slope of a line is to select any two points on the line that we will label (x_1, y_1) and (x_2, y_2). The slope can be determined by evaluating the expression

$$\text{slope} = \frac{y_2 - y_1}{x_2 - x_1}$$

Keep in mind, however, that a slope is just an indication of how fast the function's value changes: "slope = rate of change" is the way to think of it.

The important thing about a straight line is that *it has the same slope everywhere*; that is, it has a *constant slope*. That's one characteristic of a straight line: It has the same slope no matter what two points are used to calculate it, no matter where you are on the line.

Mathematical formulas that have straight-line plots can be written in a certain form that can provide certain information automatically. If a formula can be written in the form

$$y = mx + b$$

where m and b are constants (that is, numbers or expressions that contain all numbers), then an (x, y) plot of that formula will be a straight line. The constant m will be the value of the slope of the line, and the constant b will be the value of the y-intercept of the line. If we look at the equation we plotted above, according to the formula $m = 2$, so the slope of the line is 2, while $b = 1$ so the y-intercept of the plot occurs at $y = 1$.

If a formula does not have the form $y = mx + b$, it may still be possible to rearrange it algebraically so that it does have this form. If so, a plot of the rearranged equation will be a straight line.

The plots of many mathematical equations are not straight lines, but curves. One example is

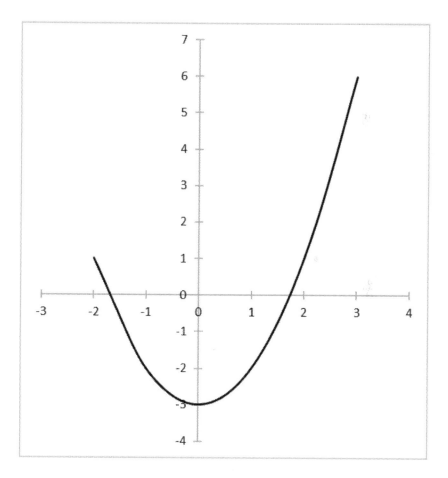

This is a curve, not a straight line. How do we know the slope of this line? Actually, there are two issues regarding the slope of a curved line. First, *the slope is different at every point of the curve*. Second, and the way we demonstrate the first issue, the slope of the curve at any given

point is represented by the tangent line that touches the curve at that point. This is illustrated here:

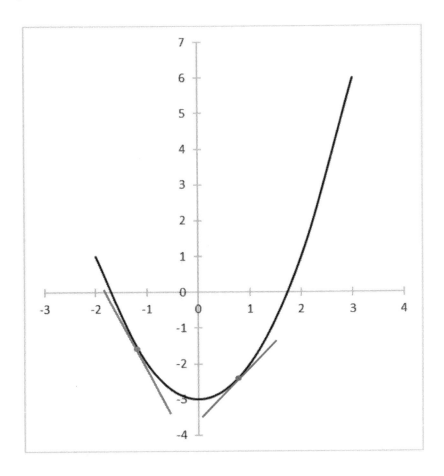

In this figure, the red line is tangent to the curve at the red dot and represents the slope of the curved line *at that point*; the blue line is tangent to the curve at the blue dot and represents the slope of the curved line at *that* point. Hopefully it's obvious that the two straight lines have different slopes, thereby demonstrating that the curved line itself has a slope that changes as the independent variable changes. By the way, the red line has a negative value for its slope; all lines that go down as you go left to right have negative slopes. On the other hand, the blue line has a positive slope, which is for lines that go up as you go left to right. Separating these two cases is a horizontal line, which has a slope of exactly 0.

How, then, do we measure or express the slope of a curve? That's where calculus comes in, and that's one reason why certain topics—including physical chemistry—use a lot of calculus in their subject matter.

2

THE DERIVATIVE

In calculus, we use a different word to represent the slope: the derivative. The important thing to remember, however, is that a derivative is nothing more than a slope—but in this case, of any line, not just straight lines.

Let us recall one way the slope is defined:

$$\text{slope} = \frac{\text{change in } y}{\text{change in } x}$$

The slope is nothing more than how much the dependent variable (y) changes when the independent variable (x) changes. The other way to express the slope is

$$\text{slope} = \frac{\Delta y}{\Delta x}$$

where the "Δ" means "change." The symbol Δ implies a finite, *non-zero* change. In fact, that's one way to estimate the slope of a curve: Select two points on the curve a certain distance away and use $\Delta y / \Delta x$ to estimate the slope of the curve. For example, we can estimate the slope of the curve using these two points:

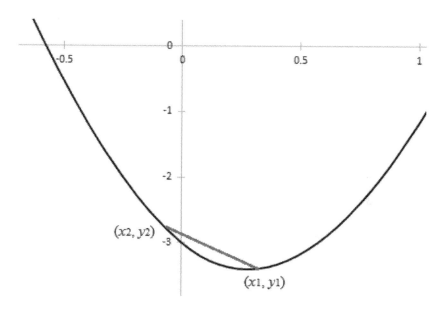

That will get us a good estimate, to be sure. We can get a better estimate, though, if we make the finite separation, the Δ, even smaller:

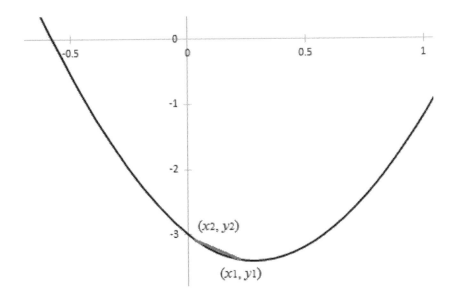

This is a better estimate of the slope of the curve in that interval, but it's still an estimate using a finite distance between the two points. In theory, we can keep getting to smaller and smaller intervals if we choose.

The best estimate—indeed, the *actual* value of the slope of the curve— can be determined if we allow the distance between the two points to shrink to zero. In mathematical terms, this is the "limit as the interval approaches zero." Because the interval is represented as Δ, the way we write this is

$$\lim_{\Delta \to 0} \frac{\Delta y}{\Delta x} = \text{actual slope of the curve}$$

As we approach this limit, the interval is no longer finite. Instead, it is infinitely small, or **infinitesimal**. At this point, to tell the difference between the finite and infinitesimal versions of the slope, for the infinitesimal version we don't use "Δ" to express the slope; we use "d" instead:

$$\text{slope} = \frac{dy}{dx} \quad \text{for an infinitesimal slope}$$

The expression "*dy/dx*," which is the value of the slope of a curve at a given point, is called the **derivative** of the curve. (Note that this is not the formal definition of "derivative," which is based on limits but is slightly more complicated. That level of complication is not needed to understand that the derivative is the slope of a line tangent to a curve.)

There are other ways to represent a derivative. If, instead of *y*, a function is expressed something like

$$F(x) = 2x^2 - 1$$

then the derivative might be written as

$$\frac{dF}{dx} \quad \text{or} \quad \frac{dF(x)}{dx} \quad \text{or even } D_x(F(x)) \quad \text{or just } D_x(F)$$

In some circumstances, the symbol for the function can be written with a prime on it, $F'(x)$, to represent the derivative.

A derivative is spoken of as, in these examples, "the derivative of *y* with respect to *x*" or "the derivative of *F* with respect to *x*" or "the derivative of *F* in terms of *x*." Note that while we're not explicitly using the word, what we really mean is "the slope of a plot of *y* [or *F*] versus *x*." The fun part comes when we use variables other than *x* and *y* in the equations we are studying. For example, in physical chemistry we study the effect of temperature on energy. In taking various measurements, for example, we may find that the total energy (*E*) depends on the temperature (*T*) according to the following equation:

$$E = C_a T^2 + C_b T + 25.0$$

where C_a and C_b are constant determined by fitting our experimental data to a quadratic equation in a spreadsheet. In this case, it would be perfectly acceptable to write the derivative of our plot as

$$\frac{dE}{dT}$$

which is interpreted as "the derivative of the energy with respect to temperature"—the slope of the plot of energy, E, versus temperature, T.

The functions that are used in physical chemistry include power functions ($x^3 + 3x^2 + ...$), trigonometric functions ($\cos \theta$, $\sin^2 4\theta$, ...), exponential functions (e^{4x}, ...), and more. Each type of function has certain rules for determining the derivative—the slope—of that function as its variable changes. The process of determining the derivative of a function is called **differentiation**. While a few examples will be presented in the next few sections, a summary of the derivatives of various functions is found near the end of this section. (Of course, most standard calculus books will have similar summaries.)

One example is a simple power function:

$$y = x^2 + 4$$

Each term on the right has the variable x raised to a power: The first term has x raised to the power of 2, and the second term can be thought of as having x raised to the 0 power. (Remember that anything raised to the 0 power is simply 1.) The rule for determining the derivative of a power function is to pull down the exponent on x and make it a multiplier, then reduce the power on x by 1; do this for every term in the equation. For the first term:

$$\overset{\text{and subtract 1}}{\overset{\frown}{\sqrt{x^2}}} \longrightarrow 2x^1 = 2x$$

For the second term, because x is treated as being raised to the 0 power, when we bring the zero down and make it a multiplier, the entire term becomes just zero. Thus, for the derivative of y with respect to x for the original equation, we have

$$\frac{dy}{dx} = 2x$$

What does this mean? First, notice that the slope itself *is a mathematical formula*. The formula is telling us that for any value of x, the slope of the plot of $x^2 + 4$ versus x is equal to $2x$. If x were 0, the slope is $2 \cdot 0$, or 0. If x is 1, then the slope is $2 \cdot 1$, or 2. If x is 4, then the slope is $2 \cdot 4$, or 8. If x is -3, then the slope is $2 \cdot (-3)$, or -6. *The derivative, the slope, changes as* x *changes.* But this is exactly what we expect of a curve— the slope of a curve changes because it's, well, a *curve*. It's not a straight line with a constant slope; it's a curved line that has a changing slope (derivative).

The derivative can be used to estimate the change in the original function. At this point, it would only be an estimate because the derivative itself changes as the independent variable (here, x) changes, but at least we can get an estimate of a function's new value if we know the derivative. Like the equation for a straight line:

$$y = mx + b$$

if we know the function's original value, F_{old}, and its slope, dF/dx, and the expected change in the independent variable Δx, then the new value of the function F_{new} can be estimated as

$$F_{new} \approx \frac{dF}{dx} \cdot \Delta x + F_{old}$$

The first term on the right is the slope of the curve times the change in x, which gives you the estimate of how much the function has changed. Add that to the original function's value and you get the new value—it's that simple! Of course, this is an estimate, because again the slope of the curve is changing over the Δx interval. But it's still a useful approximation, and it helps understand exactly what the math is saying.

The above expression for F_{new} assumes finite changes in x—and therefore in F. A lot of calculus is based on infinitesimals, however, so that begs the question: Can we express this in terms of the *infinitesimal* change in F? Of course we can, except in this case we use the expression dF to represent the infinitesimal change in the function F, which of course is based on an infinitesimal change in its variable, expressed as dx:

$$dF = \left(\frac{dF}{dx} \right) \cdot dx$$

This expression may seem trivial because algebraically, the dx terms can cancel and we're left with the identity $dF = dF$. In reality it won't be as simple as this, so the above equation is a good starting point to build on. The expression dF is called the **differential** of F, and differentials are common in parts of physical chemistry.

Remember: A derivative is just the slope of the plot of some mathematical function at a certain point.

3

PARTIAL DERIVATIVES

Many functions depend on more than one variable. One simple example is

$$F(x,y) = 3x^2y + 4y^2 - 4x$$

Here, the function F depends on two variables, x and y. How do we determine the slope, the derivative, of F?

The simple answer is "one variable at a time." When you do that, you treat all the other variables as a constant. For example, if we want to take the derivative of the above function above with respect to x, we would treat the function like

$$F(x,y) = (3y)x^2 + (4y^2) - (4)x$$

where all the constant terms, including the other variable y, are in parentheses. Now we simply take the derivative of this, treating it as a power function of x:

$$\text{derivative of } F = 2(3y)x^1 + 0 - 1(4)x^0$$

where the derivative of the second term is zero because the entire term is treated as a constant. Simplifying, we get

$$\text{derivative of } F = 6yx - 4$$

This derivative with respect to a single variable of a multi-variable function is called a **partial derivative**. Instead of using a "*d*" to represent a partial derivative, we use a stylized lowercase d, ∂, known as the **partial symbol** (it is *not* a lowercase Greek delta) so in this case the derivative of F with respect to x is represented as:

$$\frac{\partial F}{\partial x}$$

Many partial derivatives also explicitly list the other variables that are being held constant. While this is not absolutely necessary because the partial derivative construction states the variable of interest, it is technically necessary if the variables themselves depend on each other, as often happens in science. The variables held constant are listed as right subscripts outside of parentheses, like so for the equation above:

$$\left(\frac{\partial F}{\partial x} \right)_y = 6yx - 4$$

This function has another partial derivative, one with respect to y. You should satisfy yourself that this partial derivative is correct:

$$\left(\frac{\partial F}{\partial y} \right)_x = 3x^2 + 8y$$

Note that this time, the variable x is being held constant, and is listed explicitly as a right subscript outside the parentheses. Again, as with the "regular" derivative, these partial derivatives are also functions, which means that the value of the derivatives change as the values of x and y change.

If a derivative of a curve is the slope of that curve at a certain point, what do partial derivatives mean? A partial derivative is a slope of the multivariable function *in one dimension*. For example, the partial derivative $\partial F/\partial x$ is the slope of the curve (in this case, a surface) in the x dimension. The figure below shows a line tangent to the surface and parallel to the x axis: This represents the partial derivative of $F(x,y)$ with respect to just the variable x.

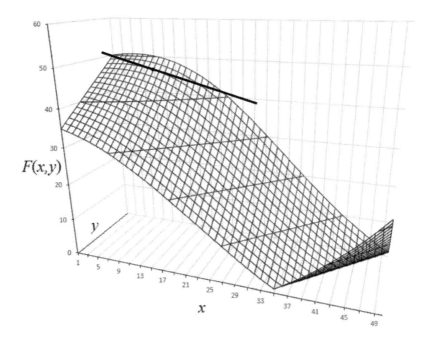

The figure below shows a similar tangent with part of the line going under the surface, but now this tangent line is parallel to the y axis. This line represents $\partial F / \partial y$.

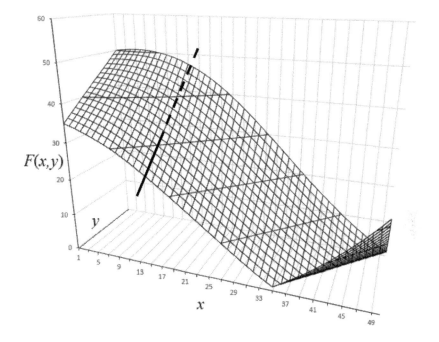

The line is parallel to the y axis (indeed, all of lines on the surface are parallel to one of the two axes), although it's difficult to see a three-dimensional plot in two dimensions.

As with the derivative in the previous section, new values of the function can be estimated if you know the derivative in either or both directions. For the x dimension, the new functional value upon a change in x is

$$F(x, y)_{\text{new}} \approx \left(\frac{\partial F}{\partial x} \right)_y \cdot \Delta x + F(x, y)_{\text{old}}$$

If the change is in the y dimension, then the new value of F is given by

$$F(x, y)_{\text{new}} \approx \left(\frac{\partial F}{\partial y} \right)_x \cdot \Delta y + F(x, y)_{\text{old}}$$

where Δy is the change in the y value. However, for this function, **both** the independent variables can change simultaneously, in which case the new value of F can be estimated by summing the changes in the two dimensions together:

$$F(x, y)_{\text{new}} \approx \left(\frac{\partial F}{\partial x} \right)_y \cdot \Delta x + \left(\frac{\partial F}{\partial y} \right)_x \cdot \Delta y + F(x, y)_{\text{old}}$$

This last expression is illustrated in the diagram below, which essentially shows the path that the value of the function takes as x changes, then y changes:

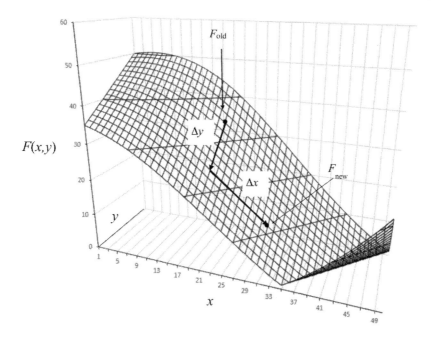

Again, F_{new} is an estimate because the changes are finite but the slopes are based on the infinitesimal slope of a curve, which is changing in both dimensions.

Partial derivatives can exist for functions in any number of variables. It helps to think of each variable as a dimension, and while we can plot a surface (which takes three dimensions, two for the variables and one for the functional value) or visualize a surface changing over time (which is, essentially, a four-dimensional representation), if a function has four or more variables it's difficult for us to visualize. But we can still think of each variable as a dimension, and if we focus on one dimension of the function at a time, keeping all the other variables constant, we've essentially reduced it into a two-dimensional plot.

4

MULTIPLE DERIVATIVES

We've seen that derivatives of functions are, themselves, functions. That means we can take the derivative of them again: We can have **multiple derivatives**.

So far, what we have taken is better called the **first derivative**. For the function

$$y = x^2 + 4$$

we found the derivative with respect to x to be

$$\frac{dy}{dx} = 2x$$

Because this derivative is itself a function of x, we can take the derivative of it with respect to x again. This would be the **second derivative**, which is represented as d^2y/dx^2, and it is determined using the same rules as when finding the first derivative of a power function:

$$\frac{d^2y}{dx^2} = 2$$

Another way to represent this is to use nested derivative constructions, as shown here:

$$\frac{d}{dx}\left(\frac{dy}{dx}\right) = 2$$

Of course, this is also a power function of x (it's x raised to the zero power), so we can take the derivative again to get the third derivative:

$$\frac{d^3y}{dx^3} = 0$$

We can continue, but there would be no point, because subsequent differentiations will simply produce zero from here on out. Some functions are differentiable only a certain number of times (as is the example here), while other functions can be differentiated over and over (like some trigonometric functions).

Previously, we introduced some different notations for a derivative. One of the more straightforward ones was the use of a prime to indicate a derivative: F' for the derivative of F. We can use more than one prime to indicate additional differentiation: F'' for the second derivative, F''' for the third derivative, and so forth. This is a convenient way to symbolize derivatives. (Note, also, that the symbol is a prime, not a quotation mark.)

A derivative with respect to time t is a common derivative. For example, basic physics uses a derivative to represent linear velocity, v, as the change in position in the x direction, represented simply by the coordinate x, with respect to time:

$$v = \frac{dx}{dt}$$

Any coordinate can be used: x, y, z, r, θ (in which case we would referring to an *angular* velocity) ... depending on the coordinate system used. Basic physics also defines acceleration, a, as the change in velocity with respect to time, so we have

$$a = \frac{dv}{dt} = \frac{d}{dt}\left(\frac{dx}{dt}\right) = \frac{d^2x}{dt^2}$$

Basic physics therefore ultimately defines acceleration as a *second* derivative of position with respect to time.

Derivatives with respect to time are relatively common in physical chemistry, so another notation is used to represent them:

$$\frac{dx}{dt} \equiv \dot{x}$$

with the dot above the variable explicitly denoting differentiation with respect to time. As you might guess, a second dot means a second derivative with respect to time:

$$\frac{d^2x}{dt^2} \equiv \ddot{x}$$

It is easy to show, then, that

$$a = \ddot{x}$$

(At least, for acceleration in the x dimension. One can also define $\ddot{y}$, $\ddot{z}$, etc., for acceleration in other dimensions.) Although this dot notation can continue to three, four... etc. dots, rarely will you see more than two dots for a derivative in physical chemistry.

Multiple derivatives can also be taken with respect to multiple variables. In this case, rather than using an exponent on the variable in the denominator of the derivative expression, each variable must be listed separately (but you can still use the exponent in the numerator). In many cases where the variables are the dimensions of the function, what we really mean is a *partial* derivative in each differentiation.

The order of derivation can be very important. In the case of multiple derivatives over multiple variables, the order of differentiation is *from the right variable to the left variable*. Thus, if we have

$$\frac{\partial^2 F(x, y)}{\partial y \, \partial x}$$

then it is assumed that the derivative with respect to x is performed first, then the derivative with respect to y. Another way to write this second derivative is

$$\frac{\partial}{\partial y} \left(\frac{\partial F(x, y)}{\partial x} \right)$$

Of course, technically we need to indicate which variables are being held constant with each partial derivative, so the most proper expression of this second derivative is

$$\left(\frac{\partial}{\partial y} \left(\frac{\partial F(x,y)}{\partial x} \right)_y \right)_x$$

As you can see, it can get very cumbersome, which is why the first representation of a multiple derivative is common.

The order of derivation can also be important because depending on the order of the variables used, *you might not get the same answer in the end*. In fact, functions that *do* give the same answer have special properties, and some of these will be useful in physical chemistry.

(Trivia time: The different ways of representing derivatives come from different well-known scientists/mathematicians:

- dy/dx and related representations were introduced by Gottfried Leibniz;
- $\dot{x}$ and similar representations were introduced by Isaac Newton;
- F' and similar representations were introduced by Joseph-Louis Lagrange;
- $D_x(F)$ and similar representations were introduced by Leonhard Euler.

Leibniz and Newton are credited with the initial development of calculus in the late 1600s, although a priority dispute between the two and their supporters continued even after their deaths. The prevalent modern perspective is that they both developed calculus independently.)

5

EXACT DIFFERENTIALS/TOTAL DERIVATIVES

Many physical quantities depend on more than one property. For example, if we were to rewrite the ideal gas law by solving for volume, what starts out as $PV = nRT$ becomes

$$V = \frac{nRT}{P}$$

We see that the volume, V, is a function of the amount n, the temperature T, and the pressure P. Thus, we can write the function as

$$V = F(n,T,P)$$

(The ideal gas law constant R is, of course, a constant.) As such, if we were to write first derivatives of the volume function, we have three variables to choose from, and so can determine three derivatives:

$$\frac{\partial F}{\partial P} \quad \text{or} \quad \frac{\partial F}{\partial T} \quad \text{or} \quad \frac{\partial F}{\partial n}$$

Here, we are using the simple form of the partial derivative. It is under-stood that the other variables not part of the derivation are being held constant. (Can you list the variables being held constant for each partial derivative?)

Each of these partial derivatives represents a slope in one dimension. The overall change in the function F, represented by the differential dF, is the sum of all these slopes multiplied by the incremental change in the variable: respectively, dP, dT, and dn:

$$dF = \left(\frac{\partial F}{\partial P}\right) dP + \left(\frac{\partial F}{\partial T}\right) dT + \left(\frac{\partial F}{\partial n}\right) dn$$

(Again, the constants for each partial derivative are omitted for clarity.) If this equation is correct for all values of the variables P, T, and n, then this expression is called an **exact differential**, sometimes called a **total differential** or a total derivative.

Most of the time in physical chemistry, exact differentials will have two terms, so they will have the general form

$$dF(x, y) = \left(\frac{\partial F}{\partial x}\right)_y dx + \left(\frac{\partial F}{\partial y}\right)_x dy$$

One of the very useful properties of exact differentials is that the following equality holds:

$$\frac{\partial^2 F(x, y)}{\partial y \, \partial x} = \frac{\partial^2 F(x, y)}{\partial x \, \partial y}$$

That is, *it does not matter what order you do the derivatives in*—you get the same final answer each time. Written in a more complicated (but ultimately useful) way, that can be stated as

$$\frac{\partial}{\partial y}\left(\frac{\partial F(x, y)}{\partial x}\right) = \frac{\partial}{\partial x}\left(\frac{\partial F(x, y)}{\partial y}\right)$$

If it turns out that $\partial F/\partial x \equiv A$ and $\partial F/\partial y \equiv B$, this simplifies into

$$\left(\frac{\partial A}{\partial y}\right)_x = \left(\frac{\partial B}{\partial x}\right)_y$$

This relationship will show up several times, especially in thermodynamics. Again, this is only guaranteed if the original differential *dF* is an exact differential.

What does this all mean? Well, recall in the discussion of partial derivatives, we showed that a function's overall change can be thought of as two separate changes, each in one dimension:

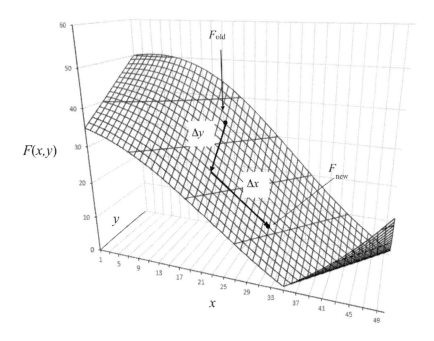

The path indicated by the dark arrows indicates that the y dimension is changing first, then the x dimension. The idea of an exact differential means that it does not matter which derivative is taken first—that is, it does not matter which dimension we pass through first. That means that this other path will get us to exactly the same place:

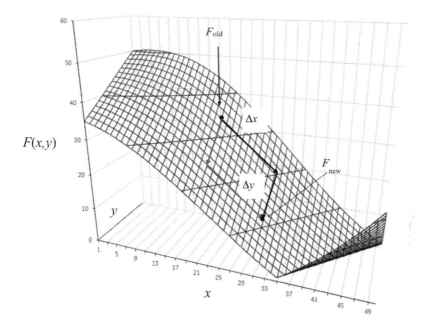

The original path has been grayed out, but it still visible. Because the change in the function F does not depend on the order of the changes— that is, because the change in F does not depend on the exact path taken—we say that the change in F is **path-independent**.

In thermodynamics, a function whose change is path-independent is called a **state function**. State functions have a certain importance in thermodynamics, and part of their importance is the mathematical properties they have by virtue of having exact differentials.

6

MAXIMA AND MINIMA OF FUNCTIONS

It does not come up very often in physical chemistry, but the idea of maximum points and minimum points ("maxima" and "minima") in functions has a few important appearances.

Consider the following graphed function:

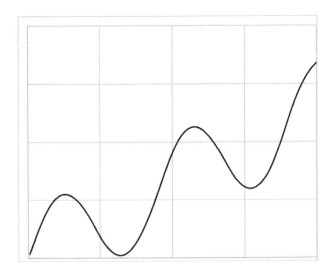

The axes aren't labeled, but that don't matter. What matters is the trend: The function goes up to some point, then down to some point, then up again, then down again, and finally up one last time.

A maximum of the function is just that—the point at which the function achieves its largest value. A minimum of a function is the opposite —the point at which the function achieves its lowest value. Ignoring the rightmost and leftmost points, hopefully it's clear that this graph has two maxima and two minima, as shown:

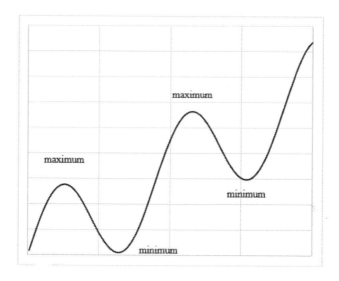

This graph shows the difference between the relative minimum/maximum and the absolute minimum/maximum: **Relative** (or **local**) minima or maxima can occur anywhere in the function, while there is only one absolute maximum and one absolute minimum in a plot. For clarity, these are labeled below:

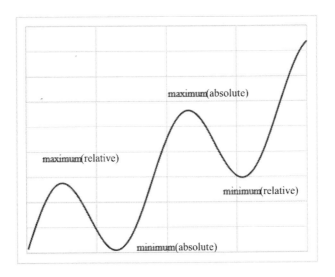

In terms of calculus and derivatives, it's very easy to understand maxima and minima: At maxima and minima, *the derivative is exactly zero*. That means that the tangent lines at the minimum and maximum points are horizontal lines, as shown:

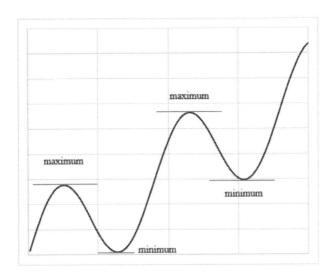

Remember, a slope is a first derivative, so another way of saying this is that the derivative of the function equals zero at any maximum or minimum:

$$\frac{\partial F}{\partial x} = 0 \quad \text{at maximum or minimum}$$

(Here we are using the partial derivative in case F is a function of more than one variable.) The idea that a maximum or minimum has a derivative that equals zero can be very useful in learning about the characteristics of certain equations in physical chemistry (or in other fields).

But if the derivative equals zero at *both* maxima and minima, how do we tell which is which? Easy—by the second derivative. Remember that if the first derivative is the change in the function—the slope—then the second derivative is the *change in the slope*. Consider the following maximum. As you go from left to right (that is, as the horizontal axis is increasing its value, which is how we conventionally read the horizontal axis), the slope on the left side of the curve is positive; the slope at the maximum is zero, and the slope on the right side of the curve is negative:

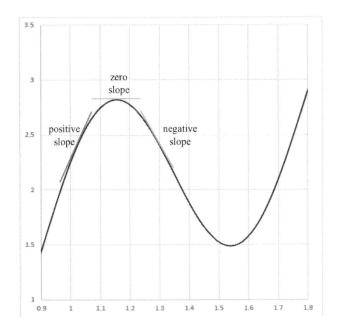

Because the slope is going from positive to zero to negative, the trend for the change in the slope is negative—think of the progression of numbers +1 → 0 → -1. Because the change is just the derivative and the slope is just a derivative, the change in the slope is the *second derivative*, and now we have our rule: *For a relative maximum, the second derivative is negative.*

For a relative minimum, it's just the opposite, as this graph shows:

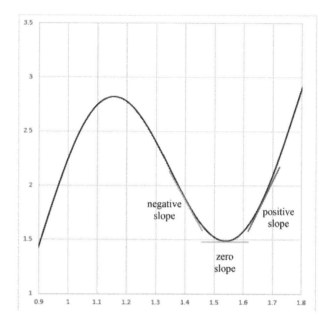

The change in the slope—that is, the second derivative—is positive, with the tangent lines going from a negative slope to a zero slope and finally a positive slope. *Relative minima have positive second derivatives.*

The second derivative cannot tell you if a relative minimum or maximum is an *absolute* minimum or maximum. The function's value at each extremum must be evaluated and compared to see which is the lowest or highest. Having said that, hopefully it is clear that calculus does give us the tools we need to be able to make those determinations.

7

THE CHAIN, PRODUCT, AND QUOTIENT RULES OF DIFFERENTIATION

Some functions are simple, as are their derivatives:

$$F(x) = \sin x$$
$$F'(x) = \cos x$$

However, some functions are more complicated in that they contain functions within functions, like this:

$$F(x) = \sin [3x^2 + 4x]$$

How do we take the derivative of functions like this?

Calculus has something called the **chain rule**. This allows us to determine derivatives for functions embedded within functions. If a function F included a function G, then the derivative of the overall function $F(G)$ is

$$[F(G)]' = F'(G)\, G'$$

That is, the derivative of the overall function is equal to the derivative of the *F* function evaluated at the original function *G* *times* the derivative of the function *G* itself. Thus, we need two steps to determine the overall derivative of the complete function. For the function *F(x)* above, the derivative of a sine function is the cosine function, so for the first part we have

$$F' = \cos [3x^2 + 4x]$$

The argument of the cosine function is the original function *G*. The function inside the original sine function is $3x^2 + 4x$; so for the second part, the derivative is simply

$$G' = 6x + 4$$

The complete derivative is the product of these two expressions:

$$[F(G)]' = (\cos [3x^2 + 4x]) (6x + 4)$$

In this example, each function is enclosed in parentheses to emphasize the two-part nature of the overall derivative. Note that the cosine is only operating on $3x^2 + 4x$; the function $6x + 4$ is multiplying the entire cosine term.

The second, third, etc., derivative can also be taken. Depending on the function, these higher derivatives can become very complicated, so it is important to be careful with the terms that are generated.

Functions can also be multiplied together (as is the first derivative that was constructed above). For example:

$$FG = 4x^2 \sin x$$

where $F = 4x^2$ and $G = \sin x$. The **product rule** of derivatives says that the derivative of a product of two functions is the derivative of the first function times the original second function *plus* the original first function times the derivative of the second:

$$[FG]' = F'G + FG'$$

That is, in the first term you keep one function the same, and in the second term you keep the other function the same. For the sample function above:

$$F' = 8x$$
$$G' = \cos x$$
$$\text{Therefore } [FG]' = 8x\sin x + 4x^2\cos x$$

As with the chain rule, the product rule can be expanded for a larger number of functions being multiplied together, and second, third, … derivatives can be taken.

Using the chain rule and the product rule, we can come up with a third rule for the quotient of two functions, called the **quotient rule**. (The actual proof of this is left for calculus texts.) Briefly, for the quotient of two functions F/G, both of which have derivatives F' and G', the derivative of the fraction F/G is

$$\left(\frac{F}{G}\right)' = \frac{G \cdot F' - F \cdot G'}{G^2}$$

While this is a bit more complicated than either the chain rule or the product rule, it does allow us to determine the derivative of a fraction of two functions.

8

DIFFERENTIAL EQUATIONS

We are used to seeing variables in equations, like $F(x) = 3x^2 + 4x$. We might also write this as $y = 3x^2 + 4x$ and understand that for every value of x, there is a corresponding value of y. We can display this on a graph, like so:

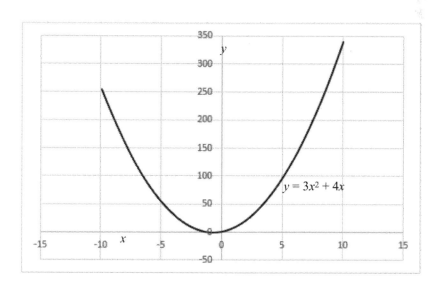

The line represents all the pairs (x,y) that satisfy the equation $y = 3x^2 + 4x$.

Variables aren't the only things that can be in equations. Derivatives can be in equations as well, and when they are, the expressions are called **differential equations**. One example of a differential equation is

$$dF = 2x$$

Solving this differential equation involves asking the question, "What function F has a derivative equal to $2x$?" There are actually several answers, but perhaps the simplest answer is $F = x^2$. When you take the derivative of this simple power function, you get $dF = 2x$. (Technically, you get $dF = 2x \, dx$.) We say that $F = x^2$ is a *solution* to the differential equation $dF = 2x$. We say that the function F *satisfies* the original differential equation.

There is some terminology associated with differential equations. An *ordinary* differential equation has a single variable. (Again, commonly x is used, but any variable is possible.) A *partial* differential equation uses partial derivatives because it depends on more than one variable, and the differential equation itself may (or may not) include differentials based on more than one variable. Differential equations may be based on a first derivative, a second derivative, etc. The highest level of derivative defines the *order* of the differential equation. If the order is 1 (that is, the differential equation only has a first derivative in it), it is called a *linear* differential equation. Such equations are relatively easy to understand and solve. If the order is greater than 1, it is a *non-linear* differential equation. Non-linear differential equations are more difficult to solve, and many are only solvable for certain specific situations.

A lot of scientific and engineering processes can be stated in terms of differential equations. Because of that, finding solutions to differential equations is a major part of science and engineering. One example is the motion of a mass attached to a wall by a spring, moving back and forth (the formal name is "harmonic oscillator"):

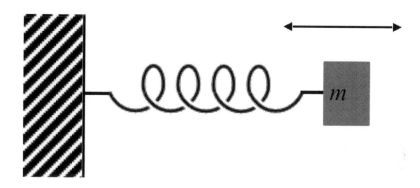

The mass m is moving back and forth, as indicated by the arrow. The total energy E of the moving mass is the sum of the kinetic energy, K, and the potential energy, V:

$$E = K + V$$

Kinetic energy is energy of motion and is given by the equation $\frac{1}{2}mv^2$, where m is the mass of the object and v is its velocity. But remember, velocity is the change in position with respect to time, so we can write v as a derivative. Using the variable x to represent the position (we could just as easily use y or z or any other variable):

$$v = \frac{dx}{dt}$$

This means we can write the kinetic energy as

$$K = \frac{1}{2}mv^2 = \frac{1}{2}m \cdot \left(\frac{dx}{dt}\right)^2$$

The potential energy of a spring is $V = \frac{1}{2}kx^2$, where k is a characteristic of the spring called its force constant. Substituting these two expressions into the equation for energy (and cleaning up the fractions a bit), we get

$$E = \frac{1}{2}m\left(\frac{dx}{dt}\right)^2 + \frac{1}{2}kx^2$$

This is a differential equation! It is a function of x as well as the second derivative of x, so it is an ordinary, non-linear differential equation with an order of 2. If we knew what function that, when substituted for x, satisfied that equation, we would know how the spring behaves, which is what the scientist or engineer wants to know.

It turns out we do know the functions that satisfy the differential equation above. They are

$$x = A\sin\left(\sqrt{\frac{k}{m}} \cdot t + \varphi\right)$$

where A is the maximum amplitude of the movement and φ (the lowercase Greek letter phi) is a constant called the *phase factor* and is determined by the initial position of the mass. The terms k and m have already been defined. That is, the mathematical expression that satisfies the differential equation we determined for our system is just a sine function. It looks like this:

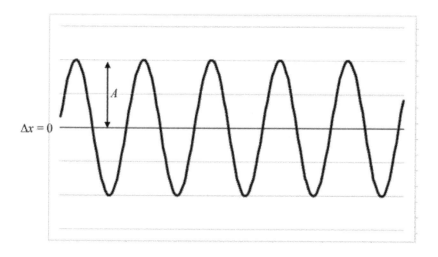

The motion is just an oscillation back and forth between a displacement of zero. This is classic harmonic motion, and we understand it by solving a differential equation.

When we know the exact function that satisfies a specific differential equation, we say that we know the **analytic solution** to the equation. We presented one such analytic solution above, to the differential equation that expresses the energy of a harmonic oscillator. In this circumstance, we were lucky—as hinted at above, not every second-order differential equation has a known analytic solution. Mathematicians can spend careers using various ways of finding (or trying to find) analytic solutions to differential equations.

If analytic solutions are so hard to find, does this mean that we can't understand systems that are described by differential equations? Not really. There are two other methods we can use:

- We can approximate an analytic solution, either by modifying the differential equation or by using a known solution—or both. In the quantum mechanics section of physical chemistry, we do both these things.
- We can approximate a numerical solution. Here we simply

substitute numbers in for the variables to see which values make the equation correct to within a certain predefined error range. With the advent of computers, analyzing differential equations numerically is relatively easily done. In fact, there are various software packages you can find (either free or for sale) that perform the numerical analyses on certain specific systems—in physical chemistry, those systems might be the quantum-mechanical behavior of atoms and molecules, or the flow of heat through a predefined volume (like a newly-designed automobile engine).

Many differential equations can look very complicated, but either analytically or approximately, scientists and engineers can use them to understand how their systems behave.

EXAMPLES OF DERIVATIVES

The following are some general expressions for derivatives that are found in physical chemistry. The chain, product, and/or quotient rules should be applied when appropriate.

Original function	Derivative	Example
Constant	0	$\dfrac{d}{dx}5 = 0$
Simple power function, x^n	$n \cdot x^{n-1}$	$\dfrac{d}{dx}x^4 = 4x^3$
Exponential function, e^x	e^x	$\dfrac{d}{dx}2e^x = 2e^x$
Trigonometric functions: $\sin x$	$\cos x$	$\dfrac{d}{dx}4\sin x = 4\cos x$
$\cos x$	$-\sin x$	$\dfrac{d}{dx}4\cos x = -4\sin x$
logarithmic functions, $\ln x$	$\dfrac{1}{x}$	$\dfrac{d}{dx}\ln x = \dfrac{1}{x}$

The following are some examples of the use of the chain rule with various functions. In some cases, the expression is left unsimplified so the reader can trace how the terms in the expression arise.

Function type	Expression	Derivative
Simple power functions	$4x^5$	$5 \cdot 4x^{5-1} = 20x^4$
	$4x^{x^2}$	$x^2 \cdot 4x^{x^2-1} \cdot 2x$
Exponential functions	e^{4x^2}	$e^{4x^2} \cdot \dfrac{d}{dx} 4x^2 = e^{4x^2} \cdot 8x$
Trigonometric functions	$\cos 4x^3$	$\dfrac{d}{dx}\cos 4x^3 = \sin 4x^3 \cdot \dfrac{d}{dx} 4x^3 = \sin 4x^3 \cdot 12x^2$
logarithmic functions	$\ln (\sin x)$	$\dfrac{1}{\sin x} \cdot \dfrac{d}{dx}\sin x = \dfrac{1}{\sin x} \cdot (-\cos x)$

10

PRACTICE

Learning is active—if you don't practice with the ideas, you won't learn. The following are some examples of taking derivatives you can use for practice. Other examples can be found online.

1.

$$\frac{d}{dx}\left(4x^2 - 7 + \frac{3}{x^2}\right) =$$

(Hint: Write the fraction as a power function using a negative exponent.)

2.

$$\frac{d}{dx}\left(e^{-3x^2/2}+4\sqrt{x}\right)=$$

(Hint: Rewrite the square root as a power function using a fractional exponent.)

3.

$$\frac{\partial}{\partial x}\left(6x^3y^2-2xy^4+5\right)=$$

4.

$$\frac{\partial}{\partial y}\left(6x^3y^2-2xy^4+5\right)=$$

5.

$$\frac{d}{dx}\left[\left(4x^2-3x\right)\left(6x^4\right)\right]=$$

6.

$$\frac{d}{dx}\left[\left(x^5 + 3x^3\right)\left(e^{-2\pi x^2/3}\right)\right] =$$

7. Find the x values where the curve $y = 3x^3 - 6x + 4$ is at a maximum or a minimum. (There may be more than one of each.)

8. Find the x values where the curve $6x^2 + 5/x$ is at a maximum or a minimum. (There may be more than one of each.)

9.

$$\frac{d}{dx}\left(\frac{4x^4 + 2x}{x^2 - 2x}\right) =$$

10.

$$\frac{d}{dx}\left(\frac{e^x}{\ln(x^3)}\right) =$$

11. Verify that the expression $F = \cos x$ is a solution of the differential equation

$$\frac{d^2F}{dx^2} + F = 0$$

PART II

INTEGRATION

11

AREA UNDER CURVES

How do you determine the area of a shape? Well, most simple shapes have simple mathematical formulas, based on their dimensions, that give you the area. For a square whose side has a length d, the area A is d^2. For a circle whose radius is r, the area A is πr^2, π ("pi") being that number 3.1415926… that we make everyone know. Other shapes have their own characteristic formulas for area.

How do we determine the area for a randomly curvy shape?

That's more difficult, because there's no given formula for the area of a random shape with curves. Initially, we need to do some approximations. Take this shape:

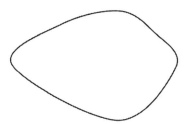

It's not very regular, so it's highly likely there's no simple formula that we can use to calculate its area. However, we can approximate its area by pretending that it can be represented by simple shapes whose areas we can calculate easily, like rectangles. Let's approximate the area of our shape like this:

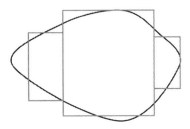

The rectangles each have a width and a height, and using them we can calculate their areas, and that will be our approximation of the area of our curvy shape. However, we can do better if we use more rectangles:

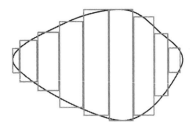

Better, but still not perfect. You probably see where we're going—we can get an exact answer if we have a very large number, indeed an infinite number, of rectangles. Of course, the larger the number of rectangles we have, the longer it takes for us to calculate the areas (even if they're as simple as "length times height"—there are *lots* of rectangles!). So there's a trade-off: accuracy on the one hand, time on the other.

It's the same thing with the area under a line of a graph. For a straight line, it's easy:

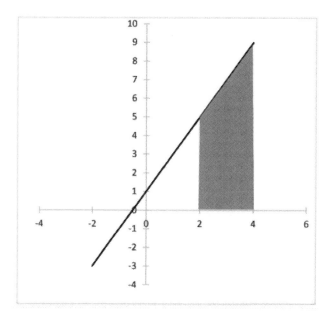

The shaded area under the line is simple to calculate because it's a trapezoid, which is a shape that has a simple formula for its area ($A = 1/2(h_1 + h_2)b$, if you're interested). However, suppose our graph is a curve, like this:

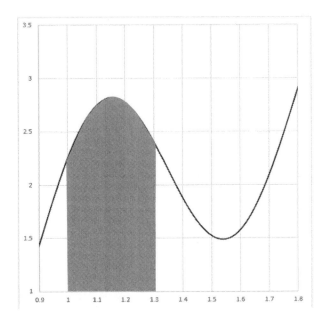

It's more difficult to determine the area.

Hey—we can use rectangles, just like we did for our curvy shape! The idea is the same: Start with a small number of rectangles to get a rough approximation of the area:

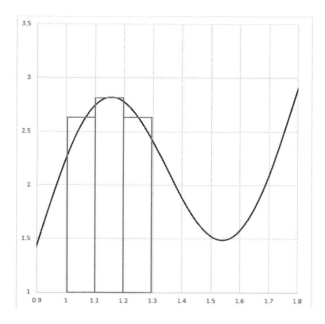

To do a better job, of course, we increase the number of rectangles:

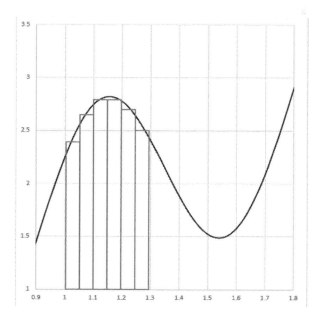

As you might expect after our discussion of the curvy shape, we can get a better and better approximation for the area under the curve if we increase the number of rectangles, ideally to infinity. But once again, it'd take a long time to calculate the area of such a large number of rectangles! The approximation of the area under a curve like this, by combining the areas of multiple shapes that mimic the area, is called a **Reimann sum**. Using the value of the function at middle of the rectangle to determine its height makes it a **middle** or **midpoint Reimann sum**.

There is one more issue here. The area of each rectangle is, of course, the width times the height. It may not be very obvious from the diagram above, but the function that is being graphed (call it $F(x)$) is playing a role: *The function's value determines the height of the rectangles*. In this case, the rectangles are drawn so their midpoint is on the graph, meaning that whatever the x value the middle of the rectangle has, the height of the rectangle is $F(x$ at the middle of the rectangle).

Again, the idea here is that we want to have an infinite number of infinitesimally-narrow rectangles to determine the area beneath the curve. The equation of the curve itself will likely be involved, because the values of the equation determine, in part, the size of the rectangles.

Finally, we should point out that area can also be negative if the curve of the function goes beneath the x axis, or the value $y = 0$. Consider this function:

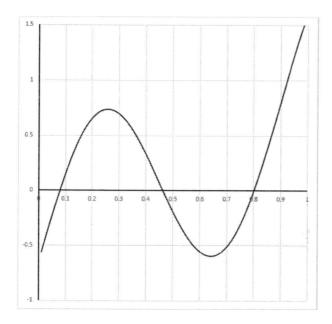

We see that the function goes above *and* below the *y*-axis (bolded). That means it has positive area *and* negative area, as shown:

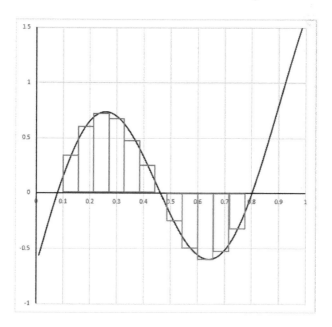

The positive and negative areas are highlighted in different colors. In this case, it's easy to see that there is more positive area than negative area, so the *total* area under the curve is positive. However, that won't always be the case, nor will it always be obvious!

12

THE INTEGRAL

In the limit of an infinite number of infinitesimally-narrow rectangles, the Reimann sum is called the **integral** of the function F. If the function is one-dimensional (and again, assuming x is the one variable), and the function ranges between $x = a$ and $x = b$, where a and b are certain values, then the integral of F is expressed by the construction

$$\int_a^b F(x)\, dx$$

The symbol looks like an elongated letter "s," because it is—it comes from the Latin word "summa," for "sum." We now refer to it as the "integral sign." The letters a and b at the bottom and top, respectively, of the integral sign are the initial and final x values of the interval of F we are considering. The "dx" is the infinitesimal of the variable x that we saw in differentiation, and it indicates what dimension (i.e. variable) the function is moving through—in this case, the only variable of the function F. (As you might expect, as with partial derivatives, we can also have partial integrals. We will cover that later.) In this context, the function F is called the **integrand**.

This expression is how we represent the area under a plot of the function F between the x values of a and b. Now the question is, how do we evaluate this expression so we can determine the area under the plot? The answer is so important that it is called the **fundamental theorem of calculus**. A simple (and possibly poor) way to state the fundamental theorem of calculus is that *the integral of a function is equal to its antiderivative*. That is:

$$\text{If } f(x) = F'(x) \qquad \text{then } F(x) = \int f(x)\,dx$$

Basically, the fundamental theorem demonstrates that the derivative and the integral are opposites of each other. But even more specifically, if we want to know the area under the curve $f(x)$ between the values of $x = a$ to $x = b$, we can determine it by

$$\int_a^b f(x)\,dx = F(b) - F(a)$$

That is, once we determine the antiderivative of $f(x)$, also known as the integral of $f(x)$, we simply evaluate *that* function values of the limits a and b and subtract. Area under the curve is now known! Such integrals area also known as **line integrals** or **path integrals** (with this second term used somewhat in physical chemistry, for certain good reasons to be covered in that class).

There are some functions whose integrals we can evaluate and some we can't. Those that we can integrate analytically are called

integrable. Again, we say we have an *analytic solution* to the integral; that is, we are able to determine a precise functional form for the integral and determine its exact value. What about equations that aren't integrable? Well, then we must go back to the idea that an integral is *an area under a curve*, and we have all sorts of ways to determine areas under any curve we can plot—we can estimate its area using rectangles, just like we did in the previous section. And, by using smaller and smaller rectangles, we can get closer and closer to the *true* area under the curve, to as many decimal points as we wish (assuming we have the time). With computers, it's relatively easy to write a simple program to determine the area under a curve using rectangles as approximations.

Let us do a simple example. We have already highlighted the area under a simple straight line as

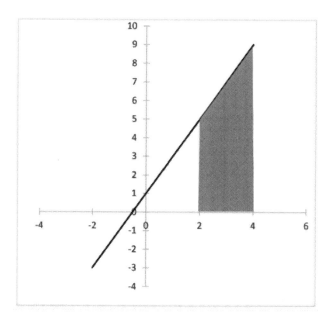

Here, our equation is $y = 2x + 1$, and the limits of the shaded area are $x = 2$ and $x = 4$. We have already argued that the area under the line can be calculated using the formula for the area of a trapezoid, which is $A =$

$1/2(h_1 + h_2)b$, where h_1 and h_2 are the heights of the sides (the y values at $x = 2$ and 4) and b is the base, which is $4 - 2 = 2$ (of whatever units). By looking at the graph (or by using the equation), we can see that h_1 and h_2 are 5 and 9, so we have

$$A = \frac{1}{2}(5+9)\cdot 2 = \frac{1}{2}\cdot 14 \cdot 2 = 14$$

of whatever (unit)^2 the function has. Now let's see doing this with an integral gets. We have to evaluate the expression

$$\int_{2}^{4}(2x+1)\,dx$$

What we need to do is determine the antiderivative of a power function. Assuming you know how to do this (and how to take antiderivatives is beyond our scope here, although some examples will be listed later), you should get

$$x^2 + x \Big|_{2}^{4}$$

where we are using the vertical bar (also called a "pipe") to indicate the upper and lower values that we need to evaluate this expression at, and subtract them. We get

$$x^2 + x \Big|_2^4 = (4^2 + 4) - (2^2 + 2)$$
$$= (16 + 4) - (4 + 2)$$
$$= 20 - 6$$
$$= 14$$

which is the same answer we got by using the trapezoid formula.

The point is, *we won't always have a simple formula for the area under a plot, especially a curvy one!* That leaves using the integral as our only option if we want to know the exact area under the curve. Again, if we can't, we can always approximate it using rectangles (or triangles or trapezoids or any other simple figure whose area formula we do know.).

Integrals are important functions in many areas of math, science, and engineering. But just like you need to remember that a derivative is just a slope, in many cases *an integral is just the area under a curve.* If the function has more than one variable, it can be more than that, but for now this is the easiest way to think of an integral.

13

INTEGRALS OF POWER FUNCTIONS

An antiderivative is just what is sounds like—the reverse of a derivative. That means that if you know how to take the derivative of a certain function, you should be able to figure out how to reverse that process and get the integral of that function.

For example, in the section on derivatives, we found that the derivative of a simple power function x^n can be expressed as nx^{n-1}:

$$\frac{\partial}{\partial x} x^n = n \cdot x^{n-1}$$

What we are doing is bringing the original power down as a multiplier, then reducing the value of the power by 1. For example,

$$\frac{\partial}{\partial x} x^5 = 5 \cdot x^{5-1} = 5x^4$$

To figure out how to take the integral of a power function, we do it backwards—increase the value of the power by 1 and divide by that new power. Thus, we have

$$\int x^n \, dx = \frac{1}{n+1} x^{n+1}$$

As an example, using the same function as an example:

$$\int x^5 \, dx = \frac{1}{5+1} x^{5+1} = \frac{1}{6} x^6$$

To verify, we can take the derivative of $\frac{1}{6}x^6$ to retrieve our original function:

$$\frac{d}{dx} \frac{1}{6} x^6 = \cancel{6} \cdot \frac{1}{\cancel{6}} x^{6-1} = x^5$$

QED.

The point is that we can try to figure out the integral of a function by knowing how to take the derivative of that type of function. This does not always guarantee that we can always determine an analytical expression for the integral of an expression, especially expressions that are combinations of functions—there is no "chain rule" for integrals like there is for derivatives, although there are some techniques for solving integrals that have more complicated integrands. Such techniques are best left to calculus classes.

14

MULTIPLE INTEGRALS

Recall that functions can have more than one variable (another way to think of variables are as "dimensions"). Recall too that it is possible to take the derivative of only one variable at a time—we called that a partial derivative. When we do a partial derivative, we treat all but one of the variables as constants, then take the derivative with respect to that one target variable.

We can do the same with integrals of functions of more than one variable. These are called **multiple integrals**, and they are represented by having two or three (or more) integral signs, one for each infinitesimal, like so:

$$\iint (x+y)\,dx\,dy$$

A construction like this is evaluated as if the integrals were *nested*, from the inside out, as if it were written like this:

$$\int\left[\int (x+y)\,dx\right]dy$$

where the first integration, the innermost one, is performed on the x variable and the second integration, the outermost one, is performed on the y variable. As with derivatives, all other variables are treated as constants when performing each one-dimensional integration.

To evaluate the multiple integral numerically, we need limits on each variable. The key here is to keep in mind that each integral sign will have limits for only the one variable. Hence, the construction

$$\int_3^8\left[\int_2^4 (x+y)\,dx\right]dy$$

implies that the inner integral is evaluated over the interval $x = 2$ to $x = 4$ and the outer integral is evaluated over the interval $y = 3$ to $y = 8$, like so:

$$\int_{y=3}^{y=8}\left[\int_{x=2}^{x=4} (x+y)\,dx\right]dy$$

As you can see, it starts getting a bit complicated to write out, so typically we use the more concise expression instead.

How do we interpret multiple integrals? Well, if a single integral represents the area under a curve, then a double integral of two variables represents the volume under a surface, like so:

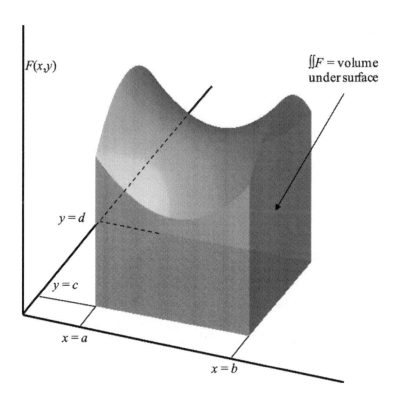

Higher multiple integrals—that is, integrals over more than two variables—have other physical interpretations that may be difficult to visualize.

There is a simplifying tactic if your expression has a certain form: **separation of variables**. In these cases, it must be possible to write the original function as a *product* of functions, each of which depends on only one of the variables. Consider the following double integral:

$$\iint e^{2x+y}\, dx\, dy$$

where the two integrations are over x and y. In this particular case, we can rewrite the function as the product of two exponential functions (taking advantage of a mathematical property of exponential functions):

$$e^{2x+y} = e^{2x}e^{y}$$

Our original double integral then can be written as

$$\iint e^{2x} \cdot e^{y} \, dx \, dy$$

What we can now do—which is something we couldn't have done with our first example—is to write as the product of *two* separate integrals, one solely for x and one solely for y:

$$\int e^{2x} \, dx \cdot \int e^{y} \, dy$$

Each individual integral can then be evaluated, and the two results simply multiplied together. In doing this, you will get the same result as you would have if you kept each variable constant and performed two integrations. However, by separating the variables, the integration should be more straightforward—in some cases, a *lot* more straightforward.

Be careful, though: You cannot separate all functions of multiple variables, even if the variables are in separate terms. For example, the integral we considered earlier

$$\iint (x+y) \, dx \, dy$$

is not separable despite the x and y variables being in different terms because we cannot write the original function as a product of a function of x and a function of y:

$$x + y \equiv F(x, y) \overset{?}{=} f(x) \cdot f(y) \; \times$$

Also, functions that have mixed terms, like $x + xy + y$, are not separable unless they can be factored in such a way that they can be expressed as $f(x)f(y)$, which is unlikely for any random function.

15

INDEFINITE AND DEFINITE INTEGRALS

There are two types of integrals that we need to be concerned with. It is important to understand the difference, because they have different applications and they can't be used in each other's places.

An **indefinite integral** is the type of integral we have been discussing so far: an antiderivative evaluated by reversing the differentiation process. For example:

$$\int x\,dx = \frac{1}{2}x^2$$

Technically, however, this is slightly incorrect. That's because the anti-derivative expression may have some constant C added to it that disappears when we take the derivative (because derivatives of constants are zero). This is called the **constant of integration**. The more proper expression is

$$\int x \, dx = \frac{1}{2}x^2 + C$$

You can verify this is correct by taking the derivative of this second function:

$$\frac{d}{dx}\left(\frac{1}{2}x^2 + C\right) = \frac{d}{dx}\left(\frac{1}{2}x^2\right) + \frac{d}{dx}(C) = x + 0 = x$$

so by taking the derivative of the expression that includes the constant C we arrive and the original integrand.

If we wanted to evaluate the antiderivative to, say, get an area under the curve $F(x) = x$, we need to evaluate this antiderivative at the limits of the interval, and here is why we've been able to get away with not using C thus far: No matter what the value of C actually is, *it cancels out when we do the subtraction*. Consider what happens when we evaluate the area between $x = 2$ and $x = 6$:

$$\int_2^6 x \, dx = \frac{1}{2}x^2 + C\Big|_2^6 = \left[\frac{1}{2}\cdot 6^2 + C\right] - \left[\frac{1}{2}\cdot 2^2 + C\right]$$

$$= \frac{1}{2}\cdot 36 + C - \frac{1}{2}\cdot 4 - C$$

$$= 18 - 2 + \cancel{C} - \cancel{C} = 16$$

Because the C constants cancel no matter what value it has, it does not affect the final answer.

These types of integrals are called *indefinite* when we don't specify the limits of integration. When we specify the limits of integration, they become **definite integrals** because we have now defined the value that the integral-with-limits represents. In many cases, the same anti-derivative expression can be used no matter what the limits are, as long as the function is defined in that interval. (Example: The plot $y = 1/x$ cannot be properly integrated between -2 and 1 because the function goes to $\pm$ as it approaches $x = 0$.) Rather than evaluating the above anti-derivative between $x = 2$ and $x = 6$, we could evaluate it between $x = -4$ and $x = 17$ if we needed to—the same antiderivative expression would give us a different, but still definite, answer.

There are other definite integrals that we use in science. These are integrals whose definite integrals have a certain, specific value *as long as the limits on the integral are the ones specified in the formula given.* For example, one common definite integral is

$$\int_0^\infty x^2 e^{-x^2}\, dx = \frac{\sqrt{\pi}}{4}$$

In this case, the value of the integral—in this case, $\sqrt{\pi}/4$—does *not* need to be evaluated at the limits listed on the integral sign: That IS the numerical value of this integral. However, it *only* applies if the limits under consideration are 0 and ∞. If they are anything else, *this value does not apply* and should not be used. This important concept applies to any predefined definite integral: The limits on the integral must be the same as in the application at hand.

Sometimes a formula for a definite integral contains some constant that appears in the integrand *and* in the solution. For example, there is a more general expression of the definite integral used in the example above:

$$\int_0^\infty x^2 e^{-ax^2}\, dx = \frac{\sqrt{\pi}}{4\sqrt{a^3}}$$

Here, "*a*" represents a constant *or collection of constants* whose presence in the integrand must be noted, because it also shows up in the solution. If, for example, $a = 3$:

$$\int_0^\infty x^2 e^{-3x^2}\, dx = \frac{\sqrt{\pi}}{4\sqrt{3^3}} = \frac{\sqrt{\pi}}{4\sqrt{27}} = \frac{1}{4}\sqrt{\frac{\pi}{27}}$$

where in the final expression we rewrote the answer as a single square-root expression. Suppose you have a mass m and a distance d that are part of your integral:

$$\int_0^\infty x^2 e^{-md^2 x^2}\, dx$$

Here, the expression md^2 is collectively a constant if in the definition of your system they are defined as such (that is, the mass is constant and the distance is constant). Because md^2 is a constant, then the entire expression md^2 is defined as the constant a and the integral becomes

$$\int_0^\infty x^2 e^{-md^2 x^2}\, dx = \frac{1}{4}\sqrt{\frac{\pi}{(md^2)^3}} = \frac{1}{4}\sqrt{\frac{\pi}{m^3 d^6}}$$

which can be simplified even further by factoring d^3 from the square-root expression. Being able to apply general expressions of definite integrals is an important skill in using those definite integrals to simplify the calculus of a problem.

16

ODD AND EVEN FUNCTIONS AND THEIR INTEGRALS

One way of simplifying the mathematics of a system is to recognize and then take advantage of any legitimate mathematical short-cuts, like those that involve some sort of symmetry. For example, a four-leaf clover can be described as a single leaf that is then rotated and repeated every 90° until you have four leaves.

Mathematical functions can be similar. One concept that uses this idea is the concept of odd and even functions. An **odd function** has the following property:

$$f(-x) = -f(x)$$

An **even function** has the following property:

$$f(-x) = f(x)$$

Graphically, it's easy to demonstrate an odd function and an even function:

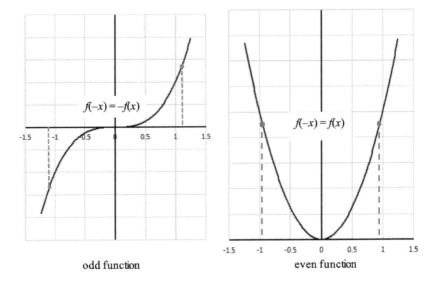

odd function   even function

Most of the time, we are considering functions that are odd or even with respect to $x = 0$ (that is, the y-axis), but that isn't necessarily the case. The definitions above would need to be shifted to the left or right, however, if the center of the interval isn't $x = 0$.

Recall that we previously mentioned that area could be positive or negative, depending on whether the curve is above or below the x-axis. This issue becomes important here. First, let us consider the odd function. If we are considering an integral of an odd function where the limits are symmetrical about the center of the function ($x = 0$ here, but again it may be elsewhere), then *the integral is exactly zero*. This is because the positive area from one side of the curve is perfectly canceled out by the negative area on the other side of the curve, as shown below:

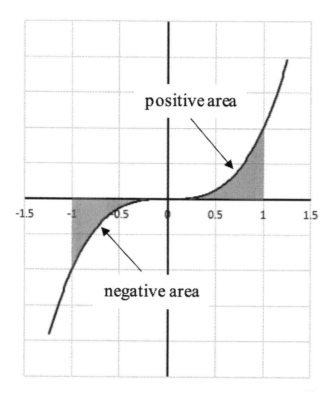

One way of writing this as an integral is

$$\int_{-a}^{+a} F(\text{odd}) \equiv 0$$

where the triple line "≡" means "is identically equal to." This is very handy, as expressions that are exactly zero can be removed from any math expression, simplifying it.

For even functions, while this is not the case, there is another simplification we can make. Because an even function has the same value on either side of the y-axis (or about whichever value of x the function is symmetrical), the area on either side of a symmetric section of the function is *exactly the same*:

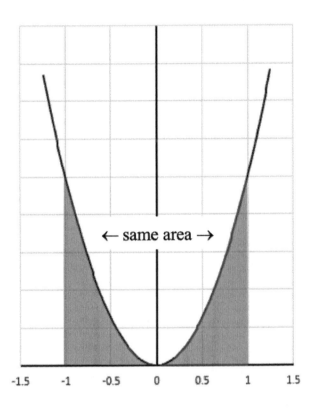

If the area under the function is an integral, we can express it as

$$\int_{-a}^{+a} F(\text{even})$$

Notice that it's not exactly zero now. What we can do, however, is split this integral into two integrals, one on each half of the interval (which, in this case, is centered at 0):

$$\int\limits_{-a}^{+a} F(\text{even}) = \int\limits_{-a}^{0} F(\text{even}) + \int\limits_{0}^{+a} F(\text{even})$$

Notice that "0" becomes one of our limits—the upper limit for the left side of the curve, and the lower limit for the right side. But if we've already demonstrated that these two areas are the same, then we conclude that these two integrals (which are areas under the curves) are the same:

$$\int\limits_{-a}^{0} F(\text{even}) = \int\limits_{0}^{+a} F(\text{even})$$

That means we can substitute 2× half the interval in place of the *entire* interval, so we have

$$\int\limits_{-a}^{+a} F(\text{even}) = 2 \times \int\limits_{0}^{+a} F(\text{even})$$

(See how the limits have changed? By the way, we could have used the −*a* to 0 integral [the left side of the curve] if we choose to; the choice is arbitrary.) Why is this useful? If you look at a table of definite integrals, you'll see that a lot of them are even functions (see below) but have defined intervals *that start at zero*. If you have a similar function that goes from some −*a* to +*a* and need to use that form of a definite integral, you can just cut the interval in half (e.g. 0 to +*a* instead of −*a* to +*a*) and include an additional multiplier of 2. Then apply the formula of the definite integral appropriately.

How do we know if a function is odd or even? Well, first, not all functions are odd or even; many functions are neither. Assuming that a function is centered about zero, one class of odd functions has all odd powers of the variable, while even functions have even powers of the variable:

$$x, x^3, 1/x, e^{-x} \ldots : \text{ odd functions}$$

$$x^2, x^4 + x^6, e^{-4x^2} \ldots : \text{ even functions}$$

When in doubt, either ask an instructor, plot it (computers are useful), or substitute $-x$ for x (or whatever the variable is) and see if you get the same function or the negative of the function—or neither, in which case it's neither odd nor every. And that means the integral needs to be figured out the regular way.

17

USING INTEGRAL TABLES

Most physical chemistry textbooks have tables of indefinite and definite integrals for reference. Any integral that is used in the text, or is expected to be used by the student in chapter exercises, should be available in those tables. If they aren't, other tables are available in other references or online.

One issue that must be addressed between instructor and student is what integrals are students expected to know and what integral will have general solutions provided, whether definite or indefinite. But whether integrals are provided or not, the ability to use a general formula for an integral is crucial to working a problem. Some may think that this is a trivial issue, but in reality it's not: Not everyone immediately understands what a general formula means and how to use it.

For example, the general formula for the integral of a power function is something we've seen before:

$$\int x^n \, dx = \frac{1}{n+1} x^{n+1}$$

Keep in mind that the variable doesn't have to be x—it can be y, z, w, Greek letters, etc. It just as easily could look like this:

$$\int \theta^{\upsilon} \, d\theta = \frac{1}{\upsilon+1} \theta^{\upsilon+1}$$

The point is, it's the *form* of the expression that's important, not the variables that are used. One thing it *can't* be is a function, because then there are chain-rule issues that must be satisfied as well. One of the necessary skills in applying these formulas is to be able to recognize the form of the integral.

Let's return to the original form of the formula. This formula tells us to:

- increase the value of the exponent by 1, and
- divide the entire expression by that new exponent ($n + 1$).

If the integral does not give us limits, technically we need to add an integration constant to it, so we can append "+ C" to the expression. If the integral *does* have limits on it, then we evaluate the expression at the top and bottom values, then subtract the bottom result from the top result. Note that the top value may be higher or lower than the bottom value, but *the order of subtraction must be the same*. Hence,

$$\int_{1}^{-2} x\,dx = \frac{1}{2}x^2\Big|_{1}^{-2}$$

$$= \frac{1}{2}(-2)^2 - \frac{1}{2}(1)^2 = 2 - 0.5 = 1.5$$

even though 1 is greater than -2. The -2 is the top number, so it is evaluated first.

Some general formulas and some of the expressions we see in physical chemistry are much more complicated than this, and this is why the ability to apply general formulas properly is an important skill. Here is one integral that we see in statistical thermodynamics:

$$\int_{n=0}^{\infty} e^{-h^2 n^2 / 8mV^{2/3}kT}\,dn$$

That's a pretty complicated expression! It does, though, have a known form and known solution, especially when you understand that h, m, V, k, and T are all constants. The form we use is

$$\int_{0}^{\infty} e^{-ax^2}\,dx = \frac{1}{2}\left(\frac{\pi}{a}\right)^{1/2}$$

This may not look like it's applicable, but again, let's focus on the form of the integral. The variable is x in the general formula but n in our integral, and both of them show up as squares in the exponent. Both functions are negative exponentials, and the limits are both zero to infinity. The challenge, however, is recognizing that a whole collection of constants in our integral represents the single constant a in our general formula. That is, in this case:

$$a \equiv \frac{h^2}{8mV^{2/3}kT}$$

Because all the symbols (including the 8) are constants, *the entire expression* is a constant and is our expression for a. That means we can substitute the entire expression for a in the solution as well:

$$\int_0^\infty e^{-ax^2}\,dx = \frac{1}{2}\left(\frac{\pi}{\dfrac{h^2}{8mV^{2/3}kT}}\right)^{1/2}$$

This compound fraction can now be simplified using the rules of algebra to get

$$\int_0^\infty e^{-ax^2}\,dx = \left(\frac{2\pi mV^{2/3}kT}{h^2}\right)^{1/2}$$

and so forth. The point is that once we can recognize that a given integral has a certain form, we can apply the solution for that form—*if* we can recognize it. That is the key skill in using integral tables.

18

USING SUBSTITUTION IN INTEGRATION

There is one technique for solving more complex integrations that is worth reviewing. In a sense, we used a simple version of it in the previous section. The concept is called **substitution**.

Consider the following integral:

$$\int_0^4 (x-5)^7 \, dx$$

While this is a simple expression, up to now our only choice for integrating would be to multiply out the binomial (seven times!) and integrate it as a power function. Straightforward to explain, but tedious to perform.

But there is another way: substitution. Suppose we define a new function, u, as the binomial $(x-5)$:

$$u \equiv (x-5)$$

We can take the derivative of both sides of this equation to get

$$du = dx$$

(You should convince yourself the expression above is correct.) What we do now is substitute u for the expression $(x - 5)$ and du for dx in our original expression:

$$\int_0^4 u^7\, du$$

This is *much* easier to integrate! We can do it simply:

$$\int_0^4 u^7\, du = \frac{1}{8} u^8 \Big|_0^4$$

What we do now is to *re-substitute* the original definition for u from above:

$$\int_0^4 u^7\, du = \frac{1}{8} u^8 \Big|_0^4 = \frac{1}{8}(x-5)^8 \Big|_0^4$$

and now the integral can be numerically evaluated easily.

One interesting point about substitution regards that final step of re-substituting the original expression back. We don't *have to* do that. But if you don't (and this is a crucial issue), *you must redefine the integration limits*. The original limits on the integral are 0 and 4—*when the variable is* x. Because we defined u as

$$u = (x - 5)$$

when x is 0, u is –5, and when x is 4, u is –1. Thus, if we want to solve the integral in terms of u, we must change the integration limits to these new values:

$$\int u^7 \, du = \frac{1}{8} u^8 \Big|_{-5 \ \leftarrow}^{-1 \ \leftarrow}$$

The arrows show that we have changed our limits to the appropriate values of u, not x. This expression can now be evaluated as-is, and no re-substitution to x is necessary. This can be a time- and effort-saving step, especially if the substitution is for a complex expression.

There are several challenges when considering integration by substitution. One challenge is to properly determine the form of the infinitesimal dx. Consider this integral:

$$\int_0^4 \sqrt{6 - \sqrt{x}} \, dx$$

We can start with the following definition:

$$u = 6 - \sqrt{x}$$

In this case, we need to solve for x, and then take the derivative to determine the proper form of the infinitesimal to substitute, as follows:

$$\sqrt{x} = 6 - u$$

$$x = (6 - u)^2 = u^2 - 12u + 36$$

Therefore: $dx = (2u - 12)du$

Also, we determine the new limits: When $x = 0$, $u = 6$, and when $x = 5$, $u = 4$. (Verify this.) Now we can substitute to get

$$\int_0^4 \sqrt{6-\sqrt{x}}\, dx \rightarrow \int_6^4 \sqrt{u} \cdot (2u-12)\, du = \int_6^4 (2u^{3/2} - 12u^{1/2})\, du$$

which can now be integrated simply as a power function in the variable u.

The second challenge with integration by substitution is to recognize when substitution is a valid and useful action. This takes a little bit of practice and sophistication. For example, for the integral

$$\int_0^{\pi/4} x\cos(5x^4)\, dx$$

substitution of $u = 5x^4$ does not work because the derivative of $5x^4$ is not x. (And no, you cannot include an additional factor of x^3 to the expression to make it work, because that would then *change the function*.) On the other hand, for the integral

$$\int_1^{10} \frac{\sin(\ln x)}{x}\, dx$$

defining $u = \ln x$ will work because the derivative of $\ln x$ is $1/x$ and our integral will become simply

$$\int \sin u\, du$$

and further evaluation is straightforward (assuming that you change your integration limits if you keep the expression in this form).

Substitution can be a useful (but tricky) method for evaluating integrals. Consult a calculus text or your instructor if you have additional questions using this technique.

19

PRACTICE

The following are some examples of taking integrals you can use for practice. Other examples can be found online. Most physical chemistry textbooks have tables of indefinite and definite integrals that you can use as reference. Other collections of integrals are available in calculus book and on the internet.

1. Evaluate

$$\int (3x^2 - 2x + 1)\, dx$$

2. Evaluate

$$\int 4e^x\, dx$$

3. Evaluate

$$\int xe^{x^2}\, dx$$

Be careful—you'll have to do the opposite of the chain rule for this.

4. Set up and evaluate the integral that gives you the area under the curve $y = 4x^2$ from 0 to 5.

5. How much more area under the curve of $y = x^2$ is there between $x = 4$ and $x = 5$ than there is under the curve between $x = 0$ and $x = 1$?

6. Evaluate the integral

$$\int_{x=3}^{5} \int_{y=0}^{3} \left(x^2 + 2y^2 \right) dy\, dx$$

7. Determine if the following functions are odd, even, or neither.

(a) $f(x) = 4x^4 + 6x^2 + 9$

(b) $f(x) = xe^{-x}$

(c) $y = 36x^2 + 12x + 3$

(d) $y = 2\sin 2x$

8. Evaluate the integral

$$\int_0^\infty x^3 e^{-k^2 x^2 / 2mT} \, dx$$

where k, m, and T are constants. Use the general formula

$$\int_0^\infty x^3 e^{-ax^2} \, dx = \frac{1}{2a^2}$$

9. Evaluate the integral

$$\int_{-\infty}^{+\infty} x^2 e^{-4x^2} \, dx$$

Use the general formula

$$\int_0^{+\infty} x^2 e^{-ax^2} \, dx = \frac{1}{4}\sqrt{\frac{\pi}{a^3}}$$

(Hint: Careful!)

10. Without doing any calculations, determine the value of the integral

$$\int_{-5}^{+5} xe^{-x}\, dx$$

(This procedure is sometimes referred to as "solving by inspection" and can be a great effort- and time-saver if used properly.)

11. Evaluate each integral using substitution. All of them can be evaluated with this technique.

(a) $\int_{0}^{2\pi} 3\cos(3x)\, dx$

(b) $\int_{-2}^{4} x(2^{x^2+3})\, dx$

(c) $\int_{2}^{5} (x+3)^3 (x-2)^2\, dx$

PART III

OTHER MATH TOPICS

20

COORDINATE SYSTEMS: CARTESIAN COORDINATES

A **coordinate system** is a method used to specify where a point is in any dimensional space. There are different coordinate systems for different dimensions, and a dimension may have more than one type of coordinate system. The one that's used may depend on the application, with certain coordinate systems having advantages over others in given applications.

The simplest coordinate system is for one dimension, which would describe a line. In many elementary schools, a *number line* is used to define a one-dimensional coordinate system:

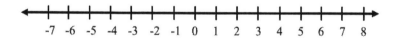

In this particular number line, the coordinates are equally-spaced (or *linear*), but they don't have to be—some coordinates are logarithmically spaced, for example. The line that labels the dimension is called an **axis** (plural **axes**), and there is usually some initial point, called the **origin**, on the axis that is usually (but not always) numbered 0 (zero). A coordinate thus specifies how far a given point is from the origin. If we were to specify the coordinate "4," then we know where our point in this space is with respect to the origin:

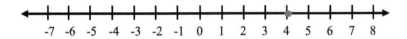

The dot indicates where we are.

In two dimensions, we need to specify two coordinates to determine the point in two dimensions:

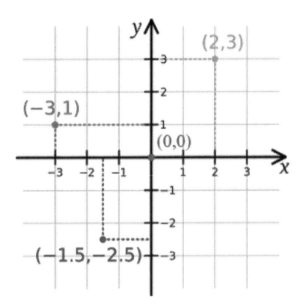

This example also demonstrates that we aren't limited whole numbers to specify a point in that space.

Coordinate systems that are formed by perpendicular axes whose unit steps (that is, the distance between whole-numbered points) are equal are called **Cartesian coordinate systems**, named after René Descartes, an early 17ᵗʰ-century French polymath who pioneered them. (In case you're wondering, the adjective derives from the Latin version of his surname, *Cartesius*.) In two dimensions, a particular point is specified by its positions along each axis, with the point itself located at the intersections of the horizontal and vertical lines crossing the respective axes. The axes are often (but not always and not required to be) referred to as the *x*-axis and *y*-axis for the horizontal and vertical dimensions, respectively, and in that order the positions are given parenthetically as an ordered pair: (2,3) thus represents the point at the intersection of a vertical line at $x = 2$ and a horizontal line at $y = 3$ (see above). In two dimensions, the *x* and *y* values are referred to as the *abscissa* and the *ordinate*, respectively.

Note that the axes for the two coordinates above are perpendicular to each other, meaning that they make a 90° angle. However, *this is not required.* It is possible to have two axes that are not perpendicular to each other, yet still provide a basis for describing two-dimensional space. For example, these two non-perpendicular axes can also be used to define any point in space:

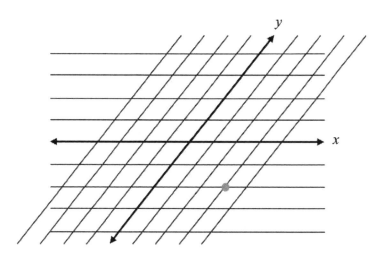

Although the axes are unlabeled, if each step were one unit, the dot would define the point (3,-2) in this axis system. Coordinate systems like this, with non-perpendicular axes, are often seen in crystals because the crystalline axes are not always perpendicular to each other.

We use three mutually-perpendicular axes to represent three-dimensional space. Again, it is usual (but not necessary) to label the axes as x, y, and z:

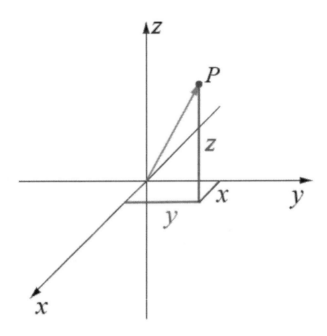

Now, any point P can be indicated by an ordered triplet (x,y,z), as shown. Because the space we live in is three-dimensional, this type of coordinate system is very common in the description of everyday events, objects, etc.

One issue in a three-dimensional coordinate system is the direction of positive and negative values. By convention, many fields use the *right-hand rule* to determine the direction of positive: Using your right hand, if the index finger points in the positive *x* direction and the middle finger, folded toward the palm, points in the positive *y* direction, then the thumb sticking up points in the positive *z* direction. This is illustrated as so:

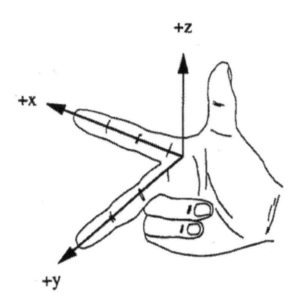

(There are other, equivalent ways of doing this, but they all should yield the same result.) The three-dimensional axes shown above conform to this convention.

As with two-dimensional space, the axes in three-dimensional space do not have to make 90° with each other.

In addition to defining points in any dimensional space, a coordinate system can be used to define a vector. If we assume that the vector starts at the origin, then the set of point(s) defines the end of the vector. From basic trigonometry, we therefore know its magnitude and its direction. For example, the vector shown below starts at the origin and goes to the point (5,4). Just knowing this, we can calculate that the vector makes an angle of 38.65° from the positive x-axis and has a magnitude of approximately 6.403 units. (If you don't recall how to determine this, you may need to review some basic algebra and trigonometry. As a start, you need to use an inverse tangent ($\tan^{-1}$) function and the Pythagorean theorem.)

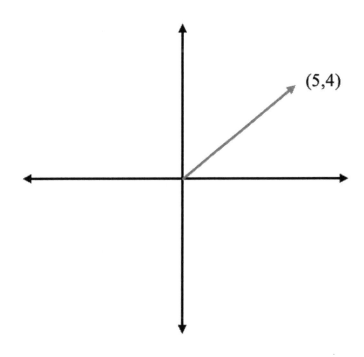

Of course, this works in three dimensions as well, with the geometry and trigonometry being only slightly more complicated.

Extending a coordinate system to vectors allows us to consider a related topic that sometime arises in science fields. Using a point in space to represent a vector is fine, but it's a *vector*. Ideally, it should be represented as vectors, not as a set of coordinates. We address this by introducing the concept of **unit vectors**, which are vectors that have a length of 1 unit (hence the name) along each of the axes. For example, in two-dimensional space, the unit vectors are commonly labeled **i** and **j** and lie on the *x*- and *y*-axes, respectively:

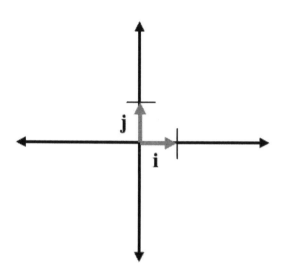

In terms of the unit vectors, any given vector **V** described by the point (*x,y*) can be written as

$$\mathbf{V} = x\mathbf{i} + y\mathbf{j}$$

Thus, if we have vector **D**:

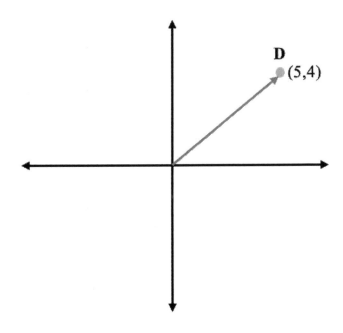

then the vector **D** is also described in terms of unit vectors as

$$\mathbf{D} = 5\mathbf{i} + 4\mathbf{j}$$

This formalism is also used to describe imaginary numbers, which is a subject we will discuss in a later section. Other fields use different notations to represent the unit vectors—$\mathbf{e}_x$ and $\mathbf{e}_y$, for example, are also common. It should be trivial to see that this is easily applied to a three-dimensional system by defining a unit vector on the z-axis, commonly labeled **k**, so that any vector **V** in three-dimensional space can also be represented as

$$\mathbf{V} = x\mathbf{i} + y\mathbf{j} + z\mathbf{k}$$

Collectively, the vectors **i**, **j**, and **k** make up what is called the **standard basis** of the space, in this case three-dimensional space. In Cartesian coordinates, the unit vectors are 90° from each other, but—as you might suspect by now—that does not always have to be the case.

21

OTHER COORDINATE SYSTEMS

As useful as Cartesian coordinates are, sometimes they are not the best way to describe a system. This is especially the case if there are any curves or circles involved.

Polar coordinates are a type of two-dimensional coordinate system that describes a point as being a certain distance **r** from an origin and a certain angle φ from some reference direction. The most common convention we see is the positive x direction as the reference direction, and positive values of φ are measured as a *counterclockwise* rotation from that direction. Hence:

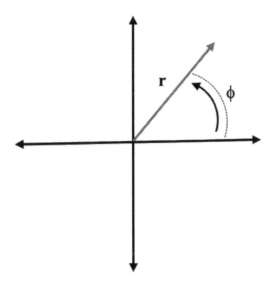

An object going in a circle, like a small mass tied to a string, can have its motion described much more easily using polar coordinates because only one variable is changing: φ. If this motion were to be described in Cartesian coordinates, two coordinates (x and y) would be constantly changing. Hence, polar coordinates simplify the treatment of rotational motion.

In extending polar coordinates into three dimensions, we have some options depending on the system or property(s) of interest. One option to express the third dimension is to use the Cartesian coordinate z. When we do that, we have **cylindrical polar coordinates**. The two previous coordinates are still defined the same way, so we have:

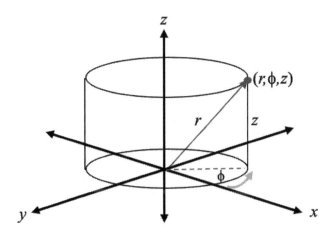

The coordinates r and φ are defined as in polar coordinates, but now there is the z coordinate also.

Another option is to use two angles and a distance. This defines **spherical polar coordinates**, and is useful for systems that are spherical in nature (like the hydrogen atom). The first two coordinates are the same as in polar coordinates, r and φ. A second angle, θ, is also defined:

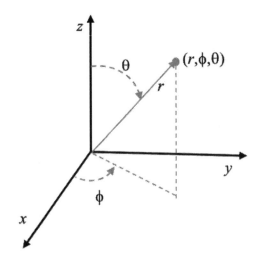

This second angle is the angle that the *r* vector makes with the positive *z* axis.

(Curiously, spherical polar coordinates are also used for terrestrial navigation, using the angles φ and θ. Except that they're not called φ and θ; they're called *longitude* and *latitude*.)

One important thing to note is that angles can be expressed in different units (this is especially important when you are using a calculator). A common unit is the *degree*, indicated by the symbol "°". Recall that there are 360° in a circle, which is a number that can be easily separated into parts to express sections of a circle—180° for a semicircle, 90° for a right angle, and so forth. However, another common unit is the *radian*, which is defined as an angular measure such that there are exactly 2π of them in a circle (so that 1 radian 57.2957...°). Again, there are benchmarks for certain fractions of a circle in terms of radians—π radians for a semicircle, $\pi/2$ radians for a right angle, and so forth.

On most (if not all) calculators, there is a setting that you must select in order to change the trigonometry mode from degree units to radian units. Most calculators automatically turn on in "deg" mode. You, the calculator user, must determine how to set it into "rad" mode if angles are being expressed in radians—or if inverse trig functions are being evaluated and the answer is expected to be in radians. When in doubt, *ask your instructor.*

Converting between the different coordinate systems is a straightforward exercise in trigonometry. Rather than derive them (again, consult a math text if you want the derivations), only the results will be presented here. Conversion of Cartesian coordinates to polar coordinates are straightforward; however, going in the opposite direction requires some care because of the properties of the trigonometric functions (sine and tangent in particular). Here we will only be concerned with conversion between Cartesian coordinates and the other systems.

To convert from two-dimensional polar coordinates to Cartesian coordinates:

Cartesian coordinate		Polar coordinates
x	$=$	$r \cos \phi$
y	$=$	$r \sin \phi$

To convert from two-dimensional Cartesian coordinates to polar coordinates (assuming angles in radians):

Polar coordinate		Cartesian coordinates
r	$=$	$\sqrt{x^2 + y^2}$
ϕ	$=$	$\arcsin\left(\dfrac{y}{r}\right)$ if $x \geq 0$ $\left[-\arcsin\left(\dfrac{y}{r}\right)\right] + \pi$ if $x < 0$

Recall that "arcsin" is the now-preferred way to denote "inverse sin" or "sin^{-1}"—especially to avoid confusion because the "-1" exponent usually means "reciprocal" (*i.e.* 1/sin).

For cylindrical polar coordinates, the conversions are exactly the same except that z in one system equals z in the other system. Thus, the tables become (again, assuming angles in radians):

Cartesian coordinate		Polar coordinates
x	$=$	$r \cos \phi$
y	$=$	$r \sin \phi$
z	$=$	z

Polar coordinate		Cartesian coordinates
r	$=$	$\sqrt{x^2 + y^2}$
ϕ	$=$	$\arcsin\left(\dfrac{y}{r}\right)$ if $x \geq 0$ $\left[-\arcsin\left(\dfrac{y}{r}\right)\right] + \pi$ if $x < 0$
z	$=$	z

For spherical polar coordinates, there are similar formulas:

Cartesian coordinate		Spherical polar coordinates
x	$=$	$r \sin \theta \cos \phi$
y	$=$	$r \sin \theta \sin \phi$
z	$=$	$r \cos \theta$

Spherical polar coordinates		Cartesian coordinates
r	$=$	$\sqrt{x^2 + y^2 + z^2}$
ϕ	$=$	$\arctan\left(\dfrac{y}{x}\right)$
θ	$=$	$\arccos\left(\dfrac{z}{r}\right) = \arccos\left(\dfrac{z}{\sqrt{x^2 + y^2 + z^2}}\right)$

Conversion between the cylindrical polar and spherical polar coordinates are omitted here.

The final aspect of converting from one coordinate system to the other is to consider how the infinitesimals change when integrating in one type of coordinate space versus another type of coordinate space. Here we will focus on the conversion between Cartesian coordinates and spherical polar coordinates, because that is the most common conversion we see in physical chemistry.

Many students miss out on the understanding that the set "$dx\ dy\ dz$" cannot simply be substituted with "$dr\ d\varphi\ d\theta$" in a three-dimensional integral (that is, "over all space"). That's because r, φ, and θ do NOT automatically substitute for x, y, and z—see the tables above! Thus, the infinitesimals dr, $d\varphi$, and $d\theta$ don't just automatically substitute for dx, dy, and/or dz.

The conversion of infinitesimals is also geometric, and is in fact based on the tables above. Rather than going through the derivations, however, only the final results will be presented here, and only for the most popular circumstances: Cartesian coordinates into (two-dimensional) polar and (three-dimensional) spherical polar coordinates.

Cartesian coordinate		Polar coordinates
In two dimensions:		
$dx\ dy$	becomes	$r\ dr\ d\phi$
In three dimensions:		
$dx\ dy\ dz$	becomes	$r^2\ \sin\theta\ dr\ d\phi\ d\theta$

In the cases where the variable r does not change, the infinitesimals simplify a bit:

Cartesian coordinate		Polar coordinates
In two dimensions, *r* **constant**:		
dx dy	becomes	$d\phi$
In three dimensions, *r* **constant**:		
dx dy dz	becomes	$\sin\theta\, d\phi\, d\theta$

Then there are the limits on the integrals. In Cartesian coordinates, the limits are either $+\infty \rightarrow -\infty$ or whatever the physical limits are to the system, if the system is confined. In polar coordinates, it's a bit trickier because as you go in a circle, you repeat the space as you complete the circle. Thus, we want to confine the integral over φ to, at most, 0 to 2π (again, in terms of radians). The limits on *r* are either 0 (*not* $-\infty$!) to $+\infty$ or to whatever limit is specified.

For spherical polar coordinates, *r* and φ have the same limits, but not θ. If we allow θ to range from 0 to 2π, we end up including the total volume of the sphere twice. (You can demonstrate this yourself: Take a round object, say a plate, and twirl it 180° around any diameter. Although you've only gone half a circle, you've already covered the entire volume of a sphere. Going a second half-circle will cover the volume a second time, overcounting it.) Therefore, the convention is that the integral over θ only ranges from 0 to π. It may sound overly complicated, but the value of using spherical polar coordinates for certain systems vastly outweighs the details of evaluating integrals in that definition of space.

22

PERPENDICULAR, NORMAL, AND ORTHOGONAL

This section is more terminology than math, but it involves three terms that are related but different. In some cases they are interchangeable, and in some cases they are not. In this discussion we will limit ourselves to Euclidean ("normal") geometry unless otherwise noted; in other circumstances these discussions may not apply.

The term **perpendicular** implies that two objects (lines, usually) meet to form a right, or 90°, angle. The two lines shown here are perpendicular:

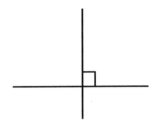

The little square in the corner is the standard notation for lines that make a 90° angle. Thus, the fundamental definition of "perpendicular" is based on a geometric representation, and a two-dimensional one at that. Of the three terms, perpendicular is the most straightforward—but the most limited.

If we expand our mix of objects, we can use a slightly different word: **normal**. If one object is a line and the other is a surface, the term "perpendicular" isn't as proper; rather, we say the objects are "normal" to each other if they make a 90° angle at the *infinitesimal point of contact*. It is clear that a line is normal to a plane if it makes a right angle with the plane:

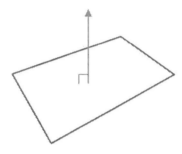

However, even in two-dimensional space the word "normal" can be used, as in the intersection of a line with a curve:

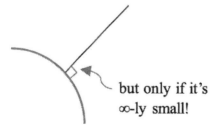

but only if it's ∞-ly small!

The straight line is perpendicular to the curve *only at the point of intersection*, so it is more proper to say that the line is *normal* to the curve at that point, rather than *perpendicular* to the curve. This is also the case with intersections involving curved surfaces. Situations like this are common in science, where a property exists perpendicular to a different property that is constantly changing, especially when those properties are vectors.

The term "orthogonal" has a broader meaning. In terms of linear algebra, two vectors or functions are **orthogonal** if their bilinear form is zero. Okay, but what's a "bilinear form?" A bilinear form is a combination of two vectors that is linear for both vectors. A *linear form* (or *linear functional*) of vectors has the following properties:

Given two vectors **A** and **B** and constant k.

The form F is linear if:

$$F(\mathbf{A} + \mathbf{B}) = F(\mathbf{A}) + F(\mathbf{B})$$
$$F(k\mathbf{A}) = k \times F(\mathbf{A})$$

Notice that **B** is not defined similarly in the last step. A bilinear form takes it one step further: It requires that the form be linear for *both* vectors separately, and also satisfy a sort of commutative property:

Given three vectors **A**, **B**, and **C**.

The function F is bilinear if:

$$F(\mathbf{A} + \mathbf{B}, \mathbf{C}) = F(\mathbf{A}, \mathbf{C}) + F(\mathbf{B}, \mathbf{C})$$
$$F(\mathbf{A}, \mathbf{B} + \mathbf{C}) = F(\mathbf{A}, \mathbf{C}) + F(\mathbf{B}, \mathbf{C})$$

Okay, but what functional, what process of the two vectors, are we talking about? For vectors, the functional is the dot product. Two vectors are orthogonal if their dot product is equal to zero:

$$\mathbf{A} \cdot \mathbf{B} = |\mathbf{A}||\mathbf{B}|\cos\theta$$

Of course, the only way the dot product is zero is if $\cos\theta = 0$, which is only true when $\theta = 90°$. So once again, we have the "perpendicular" meaning.

There is another way to determine a dot product: Multiply the respective components of the vector and add them up. This more general definition applies to vectors of more then two dimensions, where the cosine of the angle between them is harder to visualize. Hence,

$$\mathbf{A}\cdot\mathbf{B} = a_1 b_1 + a_2 b_2$$

for two dimensions, and

$$\mathbf{A}\cdot\mathbf{B} = \sum_n a_n b_n$$

for an n-dimensional vector. This explicit term-by-term sum is called a *discrete sum* because it has a finite number of terms. Any two n-dimensional vectors are orthogonal if this discrete sum is equal to zero. This is an expansion of the term "orthogonal" away from the meaning "perpendicular" because it is difficult to visualize if any two n-dimensional vectors are perpendicular to each other—but easy to see if they are orthogonal to each other (by seeing if that sum is equal to zero!).

What about, instead of vectors, we have two different functions f and g? How do we determine the inner product of two functions? The same way—multiply the functional values in the space they are defined together and add them up:

$$\sum_\tau f(\tau)\cdot g(\tau)$$

where τ (the Greek letter tau) represents the set of variables that defines the space the functions are defined in (x, (x,y,z), (r, φ, θ), and so forth). As with vectors, the two functions are orthogonal if this sum is equal to zero:

$$\sum_{\tau} f(\tau) \cdot g(\tau) = 0$$

But we have a problem. While vectors are composed of discrete coordinates, functions (at least, the well-behaved functions we usually deal with) are smoothly-varying, continuous values, not discrete points. We deal with this by making our sum finer and finer, our points closer and closer together, until they are infinitesimally close:

$$\lim_{\Delta\tau \to 0} \sum_{\Delta\tau} f(\Delta\tau) \cdot g(\Delta\tau) = 0$$

But wait a minute—*isn't this how we define an integral*? Isn't an integral an infinite sum of infinitesimally-small pieces of some function? Why, yes, it is. We therefore replace this sum over infinitesimals with an integral, like so:

$$\int_{\text{space}} f(\tau) \cdot g(\tau) \, d\tau = 0$$

This is the requirement for saying that *the functions f and g are orthogonal to each other*. The "space" limit on the integral means that it must be evaluated over the entire space of dimensions relevant to the functions.

Although we have remained true to the original definition of orthogonal—evaluating the inner product and setting it equal to zero—we can no longer use the term "perpendicular" when we are referring to functions. We therefore see that the terms "perpendicular," "normal," and "orthogonal" share certain origins, and may certainly be interchangeable in two-dimensional Cartesian coordinates, but they also have certain broader ramifications—some of which we deal with in physical chemistry.

23

AVERAGES

When dealing with lots of atoms and molecules, many macromolecular properties are essentially the average value of the individual values of each atom/molecule. Consider pressure: Pressure is caused by gas particles bouncing off the walls of the container, but because each gas particle has a different velocity, each particle exerts a different pressure. The overall pressure, the pressure we measure, is actually an average measure of the force exerted by the particles.

There are different ways to determine an average. The most straightforward one is when you have a small, finite number of values you need to average (say, exam scores). You simply add all the values and divide by the number of values:

$$\text{average} = \frac{\sum_{\#\text{ of values}} \text{value}}{\#\text{ of values}}$$

If you had exam scores of 70, 75, and 89, it is easy to show that the average is 78. In strict mathematical terms, this kind of average is called the *mean*. There is also the most probable value, mathematically called the *mode*, and the middle value of all values, called the *median*. In some special circumstances, the mode or median will be the value of interest, rather than the mean.

There is another way to determine an average (i.e. mean), using probabilities. If the probability of a certain value i is P_i, then another way to determine the average is

$$\text{average} = \frac{\text{sum of (value} \times \text{probability)}}{\text{sum of all probabilities}}$$

$$\text{average} = \frac{\sum_{\text{all values}} i \times P_i}{\sum_{\text{all values}} P_i}$$

For example, the scores on a 3-point quiz are 1, 2, 2, and 3. Thus, the probabilities of each score are

$$P_1 = 1/4$$
$$P_2 = 2/4$$
$$P_3 = 1/4$$

Using the equation above, the average is calculated as

$$\text{average} = \frac{1 \cdot \frac{1}{4} + 2 \cdot \frac{2}{4} + 3 \cdot \frac{1}{4}}{\frac{1}{4} + \frac{2}{4} + \frac{1}{4}} = \frac{\frac{1}{4} + \frac{4}{4} + \frac{3}{4}}{\frac{4}{4}} = \frac{\frac{8}{4}}{\frac{4}{4}} = \frac{2}{1} = 2$$

Although this seems like a lot of extra work when the original expression for average is easier, it has one advantage: when the probability is a function, rather than discrete values. As we saw earlier, when it's a function, the summation is replaced by an integral; thus, another useful expression for average is

$$\text{average} = \frac{\int i \cdot P_i \, di}{\int P_i \, di}$$

This definition of average has several applications in physical chemistry.

A related definition for average appears in quantum mechanics, and is dependent on an understanding of operator algebra (to be covered later). Briefly, if there is an operator A that is associated with a given measurement, then the *average value* (or *expectation value*) for that measurement, labeled <A>, is given by the expression

$$\langle A \rangle = \int_{\text{all space}} \Psi^* A \Psi \, d\tau$$

where Ψ is the state of the system. (These concepts will be covered in the quantum-mechanics section of physical chemistry.) The point to make here, however, is that this definition of an average is *a postulate of quantum mechanics*. It does not have roots in statistics or the various definitions of average presented above. As a postulate, it is unproven except in terms of the predictions it makes and how correct those predictions are when compared to reality.

24

IMAGINARY NUMBERS

A **real number** is, in simplistic terms, any number that can be represented as a length. (The formal definition of a real number is more complex than this, but need not be considered here.) For that matter, the value of *any* quantity is expressed with a real number. This includes positive and negative numbers, as well as zero. (For instance, while a length cannot be negative, a *change* in length can be negative if the absolute length gets shorter.) We have various ways of representing real numbers: as integers, as fractions, using finite decimal positions, using infinite decimal positions, with symbols (e.g., $\sqrt{2}$), even with letters (e.g., π). Real numbers have certain properties, most of which are unimportant to us. One property is important: Positive real numbers have square roots that are also real numbers. The square roots of negative numbers are not defined as being in the realm of real numbers.

However, in the 16th century, mathematicians playing with algebraic solutions of polynomials found that square, cube, etc. roots of negative numbers would appear. These roots were derisively called "imaginary" by none other than René Descartes, and the name stuck. In 1831, legendary mathematician Carl Freidrich Gauss introduced the letter i to represent **imaginary numbers**. The symbol i is one of two solutions to the algebraic equation

$$x^2 + 1 = 0$$

(The other solution is $-i$.) Effectively, $i = \sqrt{-1}$; the letter i is used to minimize confusion in certain algebraic manipulations that might seem proper but end up leading to contradictions.

Imaginary numbers are constructed by expressing the magnitude of the number and adding i to it. For example, the number $5i$ is five times the square root of 1. Imaginary numbers can also be combined with real numbers to generate a **complex number**, an example of which is 3 + $5i$. Here, the real part is 3 and the imaginary part is 5.

Complex numbers have some fascinating properties and applications that we cannot cover here. One interesting aspect of complex numbers is the **complex conjugate**, which is formed by substituting $-i$ for i wherever i appears. The complex conjugate is indicated by an asterisk (*), like so:

$$(3 + 5i)^* = 3 + 5(-i) = 3 - 5i$$

The interesting property we need to know is that *the product of a complex number and its complex conjugate is a real number.* For example:

$$(3 + 5i) \times (3 - 5i)$$
$$= 3^2 - 15i + 15i - (5i)^2$$
$$= 9 - (-25)$$
$$= 9 + 25 = 34$$

Thus, when imaginary numbers show up in physical chemistry, in some cases we will construct their complex conjugates and take the product of the two expressions as a way to generate a real number.

25

OPERATORS

In mathematics, an **operator** is an instruction to do something to a number or a function to create what may (or may not) be a new number or function. Here, we are not talking about mathematical operations like multiplication ($\times$), addition ($+$), and the like, although those functions may be part of a operator. For example, suppose we define the *Th* operator as "multiply a given function by 3." For example, if we have a function $F = 4x^2$, we can define the combination

$$Th(F) = 3 \times (4x^2) = 12x^2$$

We say that the operator *Th* operated on our function to generate a new function, which is $12x^2$. Note that we can represent the operator *Th* as "$3\times$" which may not make much mathematical sense (it's akin to an incomplete sentence in writing class), but that's because it needs something to *operate on*, which is called the **operand**.

Care must be taken with some operators, because they actually transform the function into a new function. A common operation is the derivative operation:

$$\frac{d}{dx}$$

which, if course, says "take the derivative of the operand." If we use D to represent the derivative operation, we could have

$$DF = \frac{d}{dx} 4x^2 = 8x$$

The operators we will be using will all be *linear operators*. Linear operators have the following properties:

Given two functions A and B and constant k.
The operator F is linear if:
$$F(A + B) = F(A) + F(B)$$
$$F(kA) = k \times F(A)$$

In fact, these are the same requirements we have for linear forms in our orthogonal discussion.

There is another circumstance that concerns us involving operators. A special relationship can exist between certain operators and certain functions: In some cases, invoking an operator on a function ultimately yields *a constant or collection of constants times the original function*. That is, for some operator O and function F:

$$OF = k \cdot F$$

where k is a constant or collection of constants. It is tempting to think algebraically and cancel out the function F:

$$O\,F = k \cdot F \xrightarrow{\text{???}} O = k$$

but this is not proper because the operator is an instruction to do something, not necessarily multiply by a constant! As an example, if $O = \partial/\partial x$ and $F = e^{4x}$, we have

$$\frac{\partial}{\partial x} e^{4x} = 4 \cdot e^{4x}$$

but we do not mean to suggest that

$$\frac{\partial}{\partial x} = 4 \; !$$

When you have an operator operating on an operand, you have to perform the operation and then examine the result to see if you get the original function back times a constant. If so, the operator and function make an *eigenvalue equation*, the constant you get back is called the *eigenvalue*, and the function is an *eigenfunction* of the operator. Thus, in the example above, the function e^{4x} is the eigenfunction and the eigenvalue is 4. Not all operator/function combinations result in eigenvalue equations; in fact, they are very rare, but they also have a special place in quantum mechanics.

26

LOGARITHMS

Logarithms are very useful mathematical tools. They show up in many sciences, so a review of that they are and some of their properties is presented here.

A **logarithm** is the opposite of an exponent. It answers the question, "To what power must I raise a number to get this particular answer?" More formally, the logarithm (written, represented, and spoken by the abbreviation 'log') of a number in a certain **base** b is the exponent that b needs to be raised to in order to get that number. (The base b must be a positive number and not equal to 1.) For example, the logarithm of 1000 in base 10 is 3:

$$\log_{10}(1000) = 3$$

which implies

$$10^3 = 1000 \checkmark$$

Logs are not necessarily whole numbers. For example,

$$\log_{10}(12{,}345) = 4.019149\ldots$$

If a base is not specified, it is assumed to be 10. Thus, $\log(25) = 1.39794\ldots$ implies that $10^{1.39794\cdots} = 25$. If the base is not 10, it is usually explicitly expressed:

$$\log_2(345) = 8.43045\ldots$$

One special base of logarithms is the value $2.718281828459\ldots$, which is represented by the letter e. The value of e is the base for what is called **natural logarithms**, called "natural" for reasons we will not to into here. If that base is used, the abbreviation "ln" is used, not "log."

Logarithms of numbers less than one are negative, in any base. For example,

$$\log(0.0432) = -1.364516\ldots$$

On the other hand, logarithms of negative numbers themselves are undefined. The argument of a logarithm *must* be a positive number. Although we sometimes deal with complex numbers (i.e. numbers with an imaginary component), we don't deal with logarithms of complex numbers.

No matter what the base, logarithms have some interesting properties that make them very useful in math and science. Here we tabulate some of the properties. The properties are true for any values or mathematical expressions a and b, and work in both directions.

Property	Example #1	Example #2
$a \log b = \log b^a$	$3 \log x = \log x^3$	$\log y^{3x} = 3x \cdot \log y$
$\log (ab) = \log a + \log b$	$\log (xy^2) = \log x + \log y^2$	$\log 32 + \log 3 = \log (32 \cdot 3)$
$\log (a/b) = \log a - \log b$	$\log (x^2/y^2) = \log x^2 - \log y^2$	$\log 65 - \log 5 = \log (65/5)$

Note that the first property also applies to square roots, cube roots, etc., as long as the root is expressed as a fractional exponent; i.e. $\sqrt{a} = a^{1/2}$, etc. Also, these properties can be combined. The key point to keep in mind is that *all the logs must have the same base*. Otherwise, you either can't apply them or you will have to convert all logarithms to the same base. (Consult a math text or your instructor for instructions on how to do that.)

As mentioned earlier, logarithms and exponents are opposites of each other, so can be used to cancel each other out. So,

$$\log_{10}(10^{6x^2}) = 6x^2$$

$$e^{\ln(2y)} = 2y$$

One final note: Be sure you understand how your calculator evaluates logarithms. Most calculators have "log" and "ln" for base-10 and natural logarithms, respectively. However, the order in which you hit the calculator keys to properly evaluate a log may vary by calculator model. If in doubt, consult your instructor.

27

PRACTICE

1. A two-dimensional vector starts at the origin and extends to the point (5, 8). Convert this to polar coordinates. (You may have to use some trigonometry skills that this review did not cover.)

2. A vector has the polar coordinates (5, 150°). Convert this to Cartesian coordinates, expressing your answers to two decimal places. (You may have to use some trigonometry skills that this review did not cover.)

3. A three-dimensional vector starts at the origin and extends to the point (-1, 9, 3). Convert this to spherical polar coordinates. (You may have to use some trigonometry skills that this review did not cover.)

4. A vector has the spherical polar coordinates (14.5, 83°, $\pi/4$). Convert this to Cartesian coordinates. (You may have to use some trigonometry skills that this review did not cover.)

5. Use an integral table to determine if the functions $\cos\theta$ and $\sin\theta$ are orthogonal over the integral $\theta = 0$ to 2π. That is, evaluate the integral

$$\int_0^{2\pi} \cos\theta \cdot \sin\theta \, d\theta$$

and see if it equals zero.

6. Determine if the following vectors are orthogonal.

   (a) (4, 1) and (-1, -4)
   (b) (6, -1, 2) and (2, -2, -7)
   (c) (1, 2, 3, 4, 5) and (-4, 6, -10, 8, -2)
   (d) (1, 2, 3) and (0, -6, 5)

7. An instructor give a 5-point quiz and gets these grades:

$$5, 3, 4, 4, 3, 5, 5, 1, 0, 2, 5, \text{ and } 3$$

Use two different ways to determine the average quiz score and verify that you get the same average.

8. What are the complex conjugates of each complex number?

   (a) $3 - 7i$
   (b) $7 + \pi i$
   (c) e^{-4ix}
   (d) $\sin(4i\theta/12)$

9. Evaluate the operator/function expressions given the following definitions:

Operator	Expression
$A = \dfrac{\partial}{\partial x}$	$F(x) = 4x^2$
$B = -4\pi \dfrac{\partial^2}{\partial x^2}$	$G(x) = e^{-6x}$

(a) AF
(b) AG
(c) BF
(d) BG

10. For question #9, which if any of the expressions yield eigenvalue equations, and for those that do, what is/are the eigenvalues?

11. Rearrange each expression into a single logarithm.

(a) $3 \log 2x + \log y^{1/2}$
(b) $x \log(5/y) - y \log (5/x)$
(c) $\log(1) + \log(2) + \log(3) + \log(4)$

56912185R00087

Made in the USA
Middletown, DE
25 July 2019